Crop Disease Management
Principles and Practices

Crop Disease Management
Principles and Practices

Dhananjay Dwivedi

RP
RANDOM

RANDOM PUBLICATIONS
NEW DELHI - 110 002 (INDIA)

Crop Disease Management: Principles and Practices

ISBN 978-93-51112-77-8

Published in 2014 in India by

RANDOM PUBLICATIONS

4376-A/4B, Gali Murari Lal, Ansari Road
New Delhi-110 002
Phone: +9111-43580356, 23289044
E-mail: randomexports@gmail.com; sales@randompublications.com; info@randompublications.com

Reprinted 2026

Type Setting by: Friends Media, Delhi-110089

Preface

Practices that reduce the initial levels of inoculum include selecting appropriate planting materials, destruction of crop residues, elimination of living plants that carry pathogens, and crop rotation. The selection of appropriate planting material can involve planting resistant cultivars, planting a number of mixed cultivars, using certified seed and ensuring that disease is not spread on vegetative propagating material or on equipment. Many agricultural systems are characterised by dense populations with genetic homogeneity (monocultures). Once a disease becomes established in such a plant community, it can rapidly spread to epidemic proportions. Hence the value of planting mixed cultivars, which incorporate a range of resistances. The destruction of crop residues, which can harbour many pathogens, by burying, burning or removal is an important practice performed between cropping seasons. How this is done depends on the type of crop, the type of pathogen and the size of the crop. Burying crop residues can destroy some pathogens, if ploughed in deeply enough, but some pathogens can survive, and even benefit from this process if it serves to spread them throughout the field. Burning crop debris is a common practice, especially for cereal crops, and it is a successful method of destroying many pathogens. Its success depends on the intensity of the fire. There are disadvantages to burning also, such as loss of nutrients, smoke pollutions, increased soil erosion and contributing to the greenhouse effect. Because burning can completely destroy a source of inoculum, it is commonly used in eradication campaigns against newly introduced pathogens. The elimination of living plants that carry pathogens can include both remnant or diseased crop plants and also wild plants or weeds that can act as alternative hosts between seasons. Some rusts, for example, cannot complete their life cycle in the absence of an alternate host, where they undergo sexual recombination. Removing alternate hosts can delay an outbreak, but often inoculum finds its way to crop plants via wind or another vector. Crop rotation refers to the successive planting of different crops in the same area, sometimes with a fallow, or resting, period in between crops. The most successful practice rotates crops over periods that are longer than the survival period of the pathogen, thus reducing or eliminating inoculum in the interval before a susceptible crop is planted there again. It tends not to be very effective against pathogens that survive for a long time in the soil. The rate of disease spread can be reduced by controlling the spacing of plants, humidity, moisture levels, and amount of sunlight. These factors are more easily controlled in a greenhouse environment, but some can be manipulated in the field also. In general, wider spacing of plants reduces the speed with which disease moves between them, and reducing moisture levels can inhibit infection by pathogens. The amount of time that the plants and soil are wet can be reduced by watering in the morning, spacing plants further apart, pruning or training to reduce canopy cover, and orienting rows of plants with relation to the prevailing wind. The use of shade cloths can help to control fungi that need UV light in order to sporulate.

Tillage practices have indirect effects on the spread of plant pathogens, although, some forms of inoculum can be spread extensively during tillage. Tillage can bury pathogens in the topsoil in deeper where they are less likely to cause disease. Preparation of seed beds can greatly alter the soil's texture, aeration, temperature, moisture levels and density. Tillage can also influence nutrient release in the soil and can generally benefit the crop. Farming practices have moved away from regular tillage, reducing damage to roots and spread of pathogens caused by tilling machinery. However, minimum tillage can also encourage some pathogens, such as those that feed on crop residues left on the surface of the soil. Sowing practices, such as changing time, depth and direction of sowing, and changing the density of the crop can protect plants from pathogens to which they are susceptible only at certain stages of their development. Changing the time of sowing can exploit weather conditions that are unfavourable to the pathogen, thus reducing crop losses. This might require the use of a specific cultivar that is adapted to the selected growing period, but might also be susceptible to different pathogens.

The present book is a concise and up-to-date work including all aspects of this subject. Every effort has been made in the representation manner in this book so that the students do not get confused while going through this book.

I thank all members of my team who have helped in the preparation of the book. My special thanks go to "Random Publications" who have published the book.

—Dhananjay Dwivedi

Contents

1

Infectious Plant Diseases and their Control

A systems approach to plant health includes preventing and controlling diseases. There is one thing that must be kept in mind to be successful at plant disease control. You do not control a particular disease by doing any one particular thing. Diseases are managed, prevented or controlled by groups of practices loosely grouped under the concepts of the systems approach.

Infectious organisms, called plant pathogens, are part of the environment whenever and wherever plants are in existence. Surely, we do try and practice sanitation to limit their numbers. However, we cannot achieve total "sterility" of our greenhouses. We must manage the environments of our crops so plant pathogens cannot develop and flourish.

Cultivated plants are usually more susceptible to disease than their wild relatives, partly because large numbers of the same plant are often grown closely together in pure stands. Disease-causing organisms (pathogens) often get established under these conditions. Once this happens, they may spread rapidly.

In addition, many of our valuable crop and ornamental plants are basically very susceptible to disease and would have difficulty surviving in undisturbed nature.

Finally, cultivation – be it geraniums, cabbages or oak trees – constantly disturbs nature and tends to create environmental stresses. Stresses often weaken plants and subject them to pathogenic infections.

What Causes a Plant Disease?

A plant can become diseased when it is continuously disturbed over a fairly long period of time by some factor or group of factors in its environment. Sometimes one of these disturbing factors is a living, infectious pathogen. Living pathogens include bacteria, fungi, viruses, nematodes and parasitic seed plants. Only if the disturbances are severe enough to produce noticed effects, called symptoms of poor health, do we call the plant diseased. Obviously, knowledge of normal growth habits, variety (cultivar) characteristics and normal variability of plants within a species is required for recognition of a diseased condition.

Figure: *Protect cuttings from drying out before sticking to help control stem rots.*

A System of Plant Disease Management

Employing health management systems is the most important way to proceed and succeed in growing plants profitably in a greenhouse or nursery. These are not complicated or difficult. They are simply based on the fact that a plant should not be chronically disturbed by elements of its environment with which it is has difficulty coping. Such plants are stressed and are more subject to the infection and development of infectious pathogens. Reduction of plant stress involves awareness of environmental disturbances. It also involves trying to set up and maintain a reasonably balanced environment for each and every type of plant in the greenhouse or nursery.

Beyond Stress Management

As you have no doubt realized, when you are growing dozens of different kinds of plants, it is impossible to grow each one at its precisely balanced environment. Furthermore, there are some things you cannot control as you would like.

The amount of sunlight is probably the environmental factor that gives you the most trouble. Other things happen as a result of new workers tending to the plants differently, equipment breakdowns or changes in fertilizers, growing media, etc. In other words, a bumper sticker saying "Stress Happens" might be a good one for greenhouse growers!

Pathogen Management Practices

When stress happens, plant pathogens are primed to strike. We must turn our attention to methods of dealing with pathogens. Successful pathogen management is based on accurate diagnosis of the cause, thorough knowledge of the pathogen and its disease cycle, how the host and pathogen interact with various environmental factors, practicality of possible actions and cost of various practices. The most important point in controlling a plant disease is choosing the best methods for a given situation. The best methods of control for one type of disease on a certain host may not be the best methods for another disease on the same or a different plant. Also, the use of several control measures (integrated control) is often needed. For example, several cultural control practices are often combined with a protective fungicide spray, sprench or drench program.

Infectious diseases in greenhouses or nurseries are controlled by one of four basic methods:

Exclusion. Disease-causing organisms should be excluded from production areas whenever possible. Any plant pathologist will have the practice of purchasing only pathogen-free plants, cuttings or seed high on his or her list. So what else is new? The problem is that you cannot be certain you have done this, even if you do pay extra money for "indexed" or "cultured" material. Try to do the best you can, of course. Sanitation in and around the greenhouse is an excellent pathogen exclusion measure, as well. It involves cleaning and disinfecting potting benches, soil bins, head houses, greenhouse benches, tools and equipment. Disinfecting tools and equipment is a most valuable practice to help prevent the spread of many viruses and bacterial diseases. Growing crops on raised benches is an important way to keep pathogens away from crops. Using potting media that is relatively free of plant pathogens is yet another exclusion practice. Controlling weeds that harbour insects to spread pathogens or offer survival for pathogens between crops is important.

There are other exclusion practices that serve to prevent spread of pathogens from one area to another in the greenhouse or on any particular bed or bench. The most important of these is to prevent a great number of persons from handling stock plants. Another is to avoid splashing water from plant to plant whenever possible. Protection. Growing resistant varieties is a form of protection. The genetics of the "host" protect it from pathogen infection or development. This is the most common form of protection from disease. After all, most plants do not get most diseases.

Most of us readily recognize that plants can be protected from pathogens by uniform and timely applications of disease-control chemicals (fungicides, bactericides and nematicides). Diseases caused by fungi are most commonly treated in this manner. Whereas there are many specific products that are particularly useful for particular fungal diseases, the trend of late is to use broad-spectrum products or combinations of products for general fungal disease treatment when necessary.

For leaf and flower diseases, a combination of mancozeb or chlorothalonil (contact protectants) and thiophanate-methyl (a systemic protectant) is very popular now. The new strobilurins are very useful. Flutolanil is coming into widespread use now. The group of DMI systemics are useful for powdery mildews, rusts and other diseases. For root rots, a combination of thiophanate-methyl, fludioxinal or iprodione and etridiazole or mefenoxam is commonly used. When applying fungicides, it is important to have the fungicide on the plant surface or in the soil before infection takes place. Most fungicides are protectants, not eradicants.

Prevention of pathogen infection and development conditions. Many of these practices come under the stress management practices discussed earlier. It never ceases to amaze me how tough plants are if they are not stressed. Many commonly suggested cultural practices serve to manage infectious plant diseases through the idea of preventing spread, infection and development of pathogens. These generally involve altering the air and soil environment. Keeping plants spaced or growing plants on heated floors to allow foliage to dry quickly after watering are important examples. Spacing plants ensures adequate light to counter stress. It also favors uptake of needed nutrients for good plant health.

For root health, avoid overwatering and persistent wetness of the growing media. Also, avoid excessive dryness between waterings, use well-aerated and well-drained growing media and do not allow soluble salts to build up. Injured roots favour root-rotting pathogens.

Eradication. Plant pathogens can be eliminated (eradicated) and insects that spread pathogens can be controlled by pesticides. Removal and destruction of diseased plants is a common way to help control systemic bacterial and viral diseases. These plants may be the crop you are growing, weeds or other, less important host plants. Other eradication methods include crop rotation, which "starves" out soil-invading pathogens in nurseries. As you can see from the above remarks, most of the disease-controlling systems we use today do not involve spraying or drenching with chemicals. These growing practices are effective and widely used by good growers. They take planning and follow through. This is truly a systems approach to plant health management.

Bacterial Diseases

Colibacillosis (Coliform Infections)

Problems attributed to coliform infections are often caused by strains of the *Escherichia coli* organism. There is a marked variation in severity. Problems range from severe acute infections with sudden and high mortality to mild infections of a chronic nature with low morbidity and mortality. Infections may result in a respiratory disease from air sac infection, a septicemic (blood) disease from generalized infections, an enteritis from intestinal infection or a combination of any or all of these conditions. The disease may result from a coliform infection alone as in primary infection or in combination with other disease agents as a complicating or secondary infection. Secondary infections commonly occur as a part of the classic air sac disease syndrome as a complication with *Mycoplasma gallisepticum* infections.

All ages can be affected; however, the acute septicemia in young turkeys and airsacculitis in young chickens is more common in young growing birds. High, early mortality may occur as the result of navel infections. The symptoms of this disease is caused by the *E. coli* bacteria and the toxins produced as they grow and multiply. There are many different strains or serological types within the group of *E. coli* bacteria. Many are normal inhabitants in intestinal tracts of chickens and turkeys and consequently are common organisms in the birds' environment.

A marked variation exists between different strains in their ability to cause disease. Some are severe and by themselves can cause disease while others are supposedly harmless. All degrees of pathogenicity exist between the two extremes.

The primary routes of invasion by the organism are the respiratory system and the gastrointestinal tract. Omphalitis and infections in young birds may result from entry through the unhealed navel or penetration of the egg shell prior to or during incubation. The symptoms vary with the different types of infections. In the acute septicemic form, mortality may begin suddenly and progress rapidly. Morbidity may not be apparent and birds in apparently good condition may die. However, in most cases birds are listless with ruffled feathers and indications of fever. Additional symptoms of laboured breathing, occasional coughing and rales may be apparent. Diarrhea may be evident. Mortality may be high in recently hatched chicks and poults as a result of navel infection of coliforms.

Extremely acute septicemic infection may result in sudden death with very few, if any, lesions apparent. Common lesions include dehydration, swelling and congestion of the liver and spleen and kidneys and pinpoint hemorrhages in the viscera. Fibrinous to caseous exudate in the air sacs, heart sac and on the surface of the heart, liver and lungs is a characteristic lesion. The intestines may be thickened and inflamed and may contain excess mucus and areas of hemorrhage. Navel infections, similar to those described for omphalitis may be seen in young birds. Diagnosis by laboratory means is necessary since coliform infection in its various forms may resemble and be easily confused with many other diseases. Isolation and identification of the organism by culture procedures can be accomplished relatively quickly; however, mere isolation is not sufficient to make a diagnosis. One must take into consideration the organ from which the organisms were isolated, the pathogenicity of the particular isolate and the presence of other disease agents.

Management and sanitation practices designed to reduce the number of these types of organisms in the birds' environment are necessary. In addition, reducing stress factors and other disease agents can enhance the ability of birds to defend against harmful infections. Providing adequate ventilation, good litter and range conditions, properly cleaned and disinfected equipment and facilities and high quality feed and water will improve the disease resisting status of the birds. The poultryman must always avoid overcrowding, environmental stresses like chilling or overheating and avoid vaccination or handling stress during periods when the birds are already subjected to stressful conditions. Proper egg handling, good hatchery management and implementing a good sanitation program is necessary to reduce early exposure of chicks or poults to disease organisms. It is always emphasized that problems due to one of the more pathogenic strains may occur even under ideal conditions.

The response of coliform infections to various medications is erratic and often difficult to evaluate. Under practical conditions, treatment is often disappointing. Drug sensitivity varies with the strain of *E. coli* causing the condition. Laboratory tests to determine the sensitivity to the various drugs are useful to select the most beneficial drugs. When practical, moving birds to a clean environment may be of more value than medication. For example, when outbreaks occur in growing turkeys in the brooder house, moving to range is often the best treatment.

Mycoplasmosis (CRD, Air sac, Sinusitis)

Organisms in the genus Mycoplasma are a significant cause of respiratory disease in poultry. Of the numerous species of Mycoplasma that have been isolated from domestic poultry, three are of known significance. *Mycoplasma gallisepticum* is associated with chronic respiratory disease (CRD)/air sac syndrome in chickens and turkeys and infectious sinusitis of turkeys; *Mycoplasma meleagridis* is associated with airsacculitis in turkeys; and *Mycoplasma synoviae* is the cause of infectious synovitis in chickens and turkeys.

Chronic respiratory disease (CRD), air sac syndrome and infectious sinusitis of turkeys have a common cause. CRD was first recognized as a chronic but mild respiratory disease of adult chickens. It reduced egg production but caused little or no mortality. Afterward, a condition known as "air sac disease" became a problem in young birds. It caused high mortality in some flocks. Many birds became stunted, feed efficiency was reduced, and many fowl were rejected as unfit for human consumption when processed. Infectious sinusitis in turkeys produces a sinus swelling under the eye as well as an inflammation of respiratory organs. It is a chronic disease adversely affecting growth and feed conversion. It may also cause significant mortality in young poults.

A peculiar bacterial-like organism known as *Mycoplasma gallisepticum* (MG) is common to all three conditions. CRD and sinusitis in turkeys are caused by a pure MG infections while the air sac syndrome is caused by an infection of MG in combination with *E. coli*. These conditions are triggered by acute respiratory infections such as Newcastle disease or infectious bronchitis.

Mycoplasma gallisepticum is widespread and affects many species of birds. Eradication programs have reduced the incidence in recent years. It is primarily spread through the egg. Infected hens transmit organisms and the chick or poult is infected when it hatches. Organisms may also be transmitted by direct contact with infected or carrier birds. The true CRD produces slight respiratory symptoms such as coughing, sneezing and a nasal discharge. In the air sac syndrome there is an extensive involvement of the entire respiratory system. The air sacs are often cloudy and contain large amounts of exudate. Affected birds become droopy, feed consumption decreases and there is a rapid loss of body weight.

Infectious sinusitis in turkeys occurs in two forms. When the "upper" form is present, there is only a swelling of the sinus under the eye. In the "lower" form, the lungs and air sacs are involved. The air sacs become cloudy and may contain large amounts of exudate. Both forms of the disease are usually present in the flock and frequently are present in the same bird. Diagnosis of either condition must be based on flock history, symptoms and lesions. Blood tests are useful in determining whether a flock is infected.

The answer to the MG problem in both chickens and turkeys is eradication of the disease organisms. This goal has been achieved in commercial breeding flocks with voluntary programs conducted by the National Poultry Improvement Plan (NPIP) and National Turkey Improvement Plan (NTIP). The treatment of CRD, air sac syndrome and the lower form of infectious sinusitis is not considered satisfactory. Many antibiotics have been used with varying success. Whether to give treatment is a decision that must be made on each flock based on economic factors. If treatment is attempted, give high levels of one of the broad spectrum antibiotics (Tylosin, aureomycin, terramycin, gallimycin) either in feed, drinking water or by injections. The "upper"; form of infectious sinusitis can be treated with success by injecting antibiotics into the swollen sinus cavity.

Fowl Cholera

This disease occurs throughout the country wherever poultry is produced and in recent years has become the most hazardous infectious disease of turkeys. Host range is extensive and includes chickens, turkeys, pheasants, pigeons, waterfowl, sparrows and other free-flying birds. The causative organism of fowl cholera is *Pasteurella multocida.* The organism can survive at least one month in droppings, three months in decaying carcasses and two to three months in soil. Pasteurella apparently enters tissues of the mouth and upper respiratory tract. The disease is not transmitted through the egg.

Major sources of infection include:

- Body excreta of diseased birds that contaminate soil, water, feed, etc.,
- Carcasses of birds that have died of the disease,
- Contaminated water supplies such as surface tanks, ponds, lakes and streams,
- Mechanical transmission by contaminated shoes or equipment.

Studies indicate that animals other than birds may serve as reservoirs of infection and actively spread the disease. These animals include raccoons, opossums, dogs, cats, pigs, and vermin.

The disease is seldom seen in chickens under four months of age, but is commonly seen in turkeys under this age. In the peracute form, symptoms may be absent; in the acute form some birds may die without showing symptoms, but many others are visibly ill before death. Characteristic symptoms include stupor, loss of appetite, rapid weight loss, lameness resulting from joint infection, swollen wattles, difficult breathing, watery yellowish or green diarrhea and cyanosis or darkening of the head and wattles.

Lesions may be lacking in birds dying during peracute outbreaks. When present, lesions may resemble those associated with any acute septicemic bacterial infection, often those of fowl typhoid. Typical lesions may include pinpoint hemorrhages in the mucous and serous membranes and/or abdominal fat; inflammation of the upper third of the small intestine; light, firm "parboiled" appearance of the liver; enlarged and congested spleen; creamy or solid collection of material in joints; and cheesy material in the internal ear and air spaces of the cranium of birds having twisted necks. Turkeys may have pneumonia with solidification of one or both lungs. A tentative diagnosis may be made on flock history, symptoms and postmortem lesions. A definite diagnosis depends upon isolation and identification of the organism.

Properly administered bacterins are helpful in preventing fowl cholera, particularly in turkeys. Their use must be combined with a rigid program of sanitation. In general, as it applies to the use of bacterins in turkeys, complete protection is unrealistic. Follow the manufacturer's recommendations for use of the bacterin. Vaccination in conjunction with treatment is not recommended. Sanitation practices that aid in preventing the disease are:

- Complete depopulation each year with definite breaks between older birds and their replacements,

- Implement a rodent control program,
- Dispose of dead birds properly,
- Provide safe, sanitary water,
- Clean and disinfect all houses and equipment after disposing of flock,
- Keep birds confined to the house and away from wild feral birds and animals,
- Allow contaminated ranges or yards to remain vacant for at least three months.

Although drugs usually alter the course of a fowl cholera outbreak, affected birds remain carriers and the disease has a tendency to recur when treatment is discontinued. This may necessitate prolonged treatment with drugs added to the feed and water. Sulfa drugs and broad spectrum antibiotics (Penicillin) usually control losses.

Necrotic Enteritis

Necrotic enteritis is an acute disease that produces a marked destruction of the intestinal lining of the digestive tract. Common field names (rot gut, crud and cauliflower gut) accurately describe the condition. The cause of the disease is *Clostridium perfringens*, a spore-forming, rod-shaped bacterium. Bacterial organisms and their toxins are the primary cause but coccidiosis may be a contributing factor. Most of the damage to the intestinal lining apparently is due to toxins produced by the bacterial organisms. Little is known about the spread of the disease but transmission is thought to occur by oral contact with the droppings from infected birds. Necrotic enteritis appears suddenly in the affected flock. Apparently healthy birds may become acutely depressed and die within hours. Mortality is usually between two and ten percent, but may be as high as thirty percent in severe outbreaks. Losses due to reduced growth and feed conversion may be more costly than flock mortality.

Lesions of the disease usually involve the lower half of the small intestine, but in some instances the entire length of the tract is involved. The intestine is dilated, contains dark offensive fluid and a diphtheritic cauliflower-like membrane that involves the mucosa. The lining of the intestine will have a coarse Turkish-towel appearance and portions of the lining may slough off and pass out with the intestinal contents. Diagnosis in based upon history, symptoms and findings of the characteristic lesions.

Bacitracin or virginiamycin are effective treatments administered in the feed. Bacitracin can also be given in the drinking water. Supportive vitamin treatment may enhance the effectiveness of the treatments. Preventive medication may be of value on premises where prior infections have been observed. Since coccidiosis may be a contributing factor, attention must be given to an effective coccidiosis control program.

Ulcerative Enteritis (Quail Disease)

Ulcerative enteritis is an acute or chronic infection of game birds, chickens, turkeys and other domestic fowl. Death losses may be high for young quail or pullets being raised for egg production. The cause of the disease is *Clostridium colinum*, a spore forming bacterial rod. The infection spreads by the droppings from sick or carrier birds to healthy birds. The disease organism is very resistant to disinfectants and will persist under varying environmental conditions. Birds with the acute form may die suddenly while in good flesh, whereas more chronically affected birds become listless, have ruffled feathers, whitish watery diarrhea, and develop a humped-up posture. Such birds usually die in an extremely emaciated condition. The dropping may be confused with those of birds with coccidiosis and the two diseases are often seen in the same bird. Droppings of birds with only ulcerative enteritis never contain blood.

The postmortem lesions are characteristic. The entire intestinal tract often has button-like ulcers but the lower portion is most often affected. These ulcers often perforate, resulting in local or generalized peritonitis.

Although the disease is characteristic in nature, anyone suspecting the infection should seek professional confirmation before treatment is started. Bacitracin and penicillin are the most effective drugs in the treatment and prevention of this disease. If bacitracin is used, it should be incorporated in the feed at levels up to 200 grams per ton of feed. Addition of bacitracin to the water at the rate of one teaspoon per gallon aids in controlling an outbreak of the disease. Either method of administering bacitracin will control the disease within two weeks, unless a bacitracin-resistant strain of the disease organism is present. Penicillin is also used to treat the disease if bacitracin is not effective. Raising birds on wire is an effective preventative measure. Specific drugs (bacitracin or penicillin) fed at low levels, are effective for controlling the disease in operations where the use of wire flooring is impractical.

Pullorum Disease

Pullorum disease is an acute or chronic infectious, bacterial disease affecting primarily chickens and turkeys, but most domestic and wild fowl can be infected. The cause is a bacterium named *Salmonella pullorum*. This organism is primarily egg transmitted, but transmission may occur by other means such as:

- Infected hen to egg, egg to chick, or chick to chick in incubator, chick box, brooder, or house. Survivors become infected breeders (cycle begins again),
- Mechanical transmission (carried around on clothes, shoes or equipment),
- Carrier birds (apparently healthy birds shed the disease organisms),
- Contaminated premises (from previous outbreaks).

Disease organisms may enter the bird through the respiratory (as in the incubator) or digestive systems. Most outbreaks of acute pullorum disease in chickens or turkeys result from infection while in the hatchery. Pullorum disease is highly fatal to young chicks or poults, but mature birds are more resistant. Young birds may die soon after hatching without exhibiting any observable signs. Most acute outbreaks occur in birds that are under three weeks of age. Mortality in such outbreaks may approach ninety percent if untreated. Survivors are usually stunted and unthrifty. Infection in young birds may be indicated by droopiness, ruffled feathers, a chilled appearance with birds huddling near a source of heat, laboured breathing, and presence of a white diarrhea with a "pasted-down" appearance around the vent. The white diarrhea symptom instigated the term "bacillary white diarrhea" that was commonly associated with this disease at one time. Gross lesions may be lacking in some adult birds.

Diagnosis in young birds is made by isolating the causative organism in the laboratory. In older birds, blood testing may indicate an infection but a positive diagnosis depends upon isolation and identification of the organism by laboratory procedures. Complete eradication is the only sound way to prevent pullorum disease. All hatchery supply flocks should be tested and only pullorum-free flocks used as a source of hatching eggs. Purchase chicks or poults from hatcheries that are officially recognized as "Pullorum Clean" by National Poultry Improvement Plan representatives in your state.

Treatment primarily is a salvage operation and does not prevent birds from becoming carriers. Consequently, do not keep recovered flocks for egg production. Among the drugs used to treat pullorum disease are furazolidone, gentamycin sulfate, and sulfa drugs (sulfadimethoxine, sulfamethazine, and sulfamerazine).

Fowl Typhoid

Fowl typhoid is an infectious, contagious bacterial disease that is usually acute but sometimes chronic. It affects most domestic and wild fowl including chickens, turkeys, ducks, pigeons, pheasants and other game birds. It must not be confused with typhoid fever in humans that is caused by a distinctly different organism. The cause in the bacterium, *Salmonella gallinarum*. Methods of transmission are the same as for pullorum disease, including egg transmission. However, mechanical transmission is more prevalent with this disease than with pullorum disease. Any age bird can be infected, but the disease primarily occurs in young adults (usually those older than twelve weeks of age). Mortality varies from less than one to about forty percent, but higher mortality has been observed. Signs include sudden or sporadic mortality, listlessness, green or yellow diarrhea (accompanied with pasting of the vent feathers), loss of appetite, increased thirst and a pale, anaemic appearance of comb and wattles.

A tentative diagnosis may be made from the history, signs and lesions but final diagnosis must be based on isolation and identification of the causative organism. Lesions observed at necropsy help verify a diagnosis of fowl typhoid. The lesions include an enlarged and mottled spleen, enlarged liver (coloured yellow or greenish brown), small pinpoint hemorrhages in muscles and fat surrounding internal organs, and a slimy inflammation of the front third of the small intestine. Small, white plaque-like areas are visible through the walls of the intestine is suggestive of fowl typhoid in turkeys. Blood tests used to detect pullorum reactors are also used to identify fowl typhoid birds. Prevention and control depend heavily upon basic disease prevention practices including the hatching chicks from disease-free flocks (as determined by pullorum testing), practicing strict sanitation on the farm, providing clean feed and water, and proper disposal of all dead birds as approved by the state animal health agency. The causative organism can live outside the bird body for at least six months, thus requiring extra management precautions to break the disease cycle. Following an outbreak, thoroughly clean and disinfect the premises. When feasible, practice range rotation and other special precautions to prevent the carryover of infection to the following flock.

Drugs cannot be depended upon as a means of typhoid prevention and are not recommended for that purpose. Infected birds may be salvaged using the same drugs as used to salvage pullorum infected birds.

Botulism

Botulism is a disease caused by the ingestion of a toxin produced by the *Clostridium botulinum* bacterium. All domestic fowl and most wild birds are susceptible to the toxin's effects. Many human deaths have also been attributed to the consumption of food or water containing the toxin.

Botulism is not a bacterial infection, but a condition produced by a byproduct of the bacteria's growth. The organism is common in nature and is widely dispersed in soils. Ingestion of the organism is not harmful. It becomes dangerous only when conditions are favorable for its growth and subsequent toxin formation. The organism grows best under high humidity and relatively high temperature and in an environment containing decaying organic material (plant or animal). The organism requires an environment in which all atmospheric oxygen is eliminated. The organism cannot multiply in the presence of air. Stagnant pools or damp areas with buried decaying matter are danger areas for toxin development. Botulism results after the decaying animal or plant material containing the toxin is consumed. Decaying carcasses are a frequent source of the toxin, as are many insects feeding in the same tissue. The insects may contain enough toxin to cause the disease in any bird that ingests it. Since the toxin is water soluble, water sources may become contaminated and provide a reservoir for the disease.

The toxin is one of the most potent discovered by scientists. The toxin is relatively heat stable but may be destroyed by boiling. There are different types of the toxin; types A and C cause the disease in birds while type B frequently produces the disease in man.

Weakness is generally the first sign of the illness and is followed by progressive flaccid paralysis of the legs, wings and neck. When neck muscles are affected the head hangs limp, thus causing a condition referred to as "limberneck". Affected birds may have a peculiar trembling, loose feathers that are pulled out easily and dull partly closed eyes. Some birds (turkey) do not develop loose feathers or limberneck symptoms. Because of the paralysis, birds are unable to swallow and mucous accumulates in the mouth. Fatally affected birds may lie in a profound coma appearing lifeless for several hours before death. Significant lesions are not usually observed in affected birds. Examining digestive contents may reveal insects, decomposed animal or vegetable material or other matter suggesting that the birds have consumed the toxin. A tentative diagnosis can be made from the history, symptoms and post-mortem findings. As an aid to diagnosis, sick birds may be given water into the crop, kept in a cool environment and treated intravenously with antitoxin. Recovery of a large percentage of the affected birds would substantiate diagnosis.

Prevention should be aimed at eliminating sources of toxin production and preventing access of birds to such materials. These practices include prompt removal of all dead animals from houses and pens, debeaking the birds, controlling fly and insect populations and avoiding access to decaying organic material. Contaminated water supplies are particularly dangerous.

If the disease strikes, locate and remove the source of the toxin and separate all visibly affected birds from the flock for treatment. Place sick birds in a cool shaded area and give fresh water into the crop, twice daily. Mild laxatives may be used for birds that have been exposed but do not show disease symptoms. Epsom salts (one pound per 100 birds) may be mixed into feed. Adding a level teaspoonful of Epsom salts in one ounce of water and placing in the crops of sick birds has been beneficial in many instances. Antitoxin therapy is indicated only in birds that have high individual value since the antitoxin is difficult to obtain and is expensive.

Infectious Coryza

Infectious coryza is a specific respiratory disease in chickens that occurs most often in semi-mature or adult birds. Infection may result in a slow-spreading, chronic disease that affects only a small number of birds at one time, or in a rapid spreading disease with a higher percentage of birds being affected. The occurrence of infectious coryza is not widespread and the incidence is relatively low.

The disease is caused by a bacterium known as *Hemophilus gallinarum*. Outbreaks usually result from the introduction of infected or carrier birds into a flock. Transmission of the infection occurs by direct contact, airborne infection by dust or respiratory discharge droplets and drinking water contaminated by infective nasal exudate. Susceptible birds usually develop symptoms within three days after exposure to the disease. Recovered individuals may appear normal but remain carriers of the organism for long periods. Once a flock is infected, all birds must be considered as carriers.

The most characteristic symptoms of infectious coryza include edematous swelling of the face around the eyes and wattles, nasal discharge and swollen sinuses. Watery discharge from the eyes frequently results in the lids adhering together. Vision may be affected because of the swelling. The disease results in a decrease in feed and water consumption and an increase in the number of cull birds. An adverse effect on egg production usually occurs in proportion to the number of affected birds. Diagnosis can be confirmed only by isolation and identification of the causative organism. The organism, *Hemophilus gallinarum*, is extremely fastidious and often difficult to isolate.

Prevention is the only sound approach in controlling infectious coryza. It usually can be prevented by management programs that eliminate contact between susceptible and infected birds. It requires only separating affected or carrier birds from the susceptible population. In order to prevent the infection, introduce started or adult birds only from sources known to be free of the infection. If infection occurs, complete depopulation followed by thorough cleaning/disinfecting is the only means for eliminating the disease. A number of drugs are effective for treating the symptoms of the disease although the disease is never completely eliminated. Sulfadimethoxine or sulfathiazole in the feed or water or erythromycin administered in the drinking water can reduce the symptoms of this disease.

Omphalitis

Omphalitis may be defined technically as an inflammation of the navel. As commonly used, the term refers to improper closure of the navel with subsequent bacterial infection (navel ill; mushy chick disease). Apparently, most problems result from mixed bacterial infections including the common coliforms and various species belonging to the genera Staphylococcus, Streptococcus, Proteus, and others. Omphalitis can usually be traced to faulty incubation, poor hatchery sanitation or chilling/overheating soon after hatching (such as in transit). The significance of isolating one of the bacterial species mentioned above is complicated in that many of the same species can be isolated from the yolks of supposedly normal birds immediately after hatching.

Omphalitis occurs during the first few days of life, so it cannot be considered transmissible from bird to bird. It is transmitted from unsanitary equipment in the hatchery to newly hatched birds having unhealed navels. Affected chicks usually appear drowsy or droopy with the down being "puffed up". They also generally appear to be of inferior quality and show a lack of uniformity. Many individuals stand near the heat source and are indifferent to feed or water.

Diarrhea sometimes occurs. Mortality usually begins within 24 hours and peaks by five to seven days. Characteristic lesions are poorly healed navels, subcutaneous edema, bluish colour of the abdominal muscles around the navel and unabsorbed yolk material that often has a putrid odor. Often yolks are ruptured and peritonitis is common. A tentative diagnosis can be made on the basis of history and lesions. The presence of mixed bacterial infections and absence of any specific disease-producing agent is used for confirming the diagnosis. Good management and sanitation procedures in the hatchery and during the first few days following hatching are the only sure ways to prevent omphalitis. Broad spectrum antibiotics help reduce mortality and stunting in affected groups, but they do not replace sanitation.

Erysipelas

Erysipelas is a bacterial disease caused by *Erysipelothrix insidiosa*. The disease affects several species of birds including chickens, ducks and geese, but the fowl in which it has been of primary importance is the turkey. Man is susceptible to infection and may contract the disease from infected turkeys. Since this organism is pathogenic for man, care should be taken when handling infected birds or tissues. Erysipelas in turkeys occurs most often during the fall and winter months and usually affects birds that are four to seven months of age, although any age bird is susceptible. Incidence has often been reported to be higher in males than in females, possibly because fighting males receive numerous skin abrasions that serve as portals of entry for the bacteria. In some instances the incidence is higher in hens than toms because of artificial insemination techniques that provide a means of transmission.

The organism may survive for long periods in the soil and most outbreaks are thought to originate from contaminated soil or premises. Sheep, swine and rodents may be carriers of the disease organisms. Recurrence of the disease on a premise is common. Predisposing or aggravating factors include over-crowding damp or inclement weather and poor sanitation and range management. The first indication of the disease may be the discovery of several dead birds. Usually several morbid birds can be found; however, most affected birds are visibly sick for only a short period before death. Symptoms are typical of a septicemic disease and include a general weakness, listlessness, lack of appetite and sometimes a yellowish or greenish diarrhea. Occasionally, the snood of toms may be turgid, swollen and purple. Some birds may be found lame with swollen leg joints due to localization of the infection. In breeding flocks, this disease occasionally is associated with decreased fertility and hatchability. Daily morbidity and mortality usually are low; however, in untreated flocks mortality may persist for some time and become excessive.

The most characteristic lesions are small or diffuse hemorrhages located in almost any tissue or organ. Such hemorrhages are commonly observed in the muscles, heart, liver, spleen, fat and other tissues of the body cavities. Hemorrhagic conditions of skin may result in purple blotches. The liver and spleen are usually enlarged, congested and occasionally contain necrotic foci. Enteritis or inflammation of the intestinal tract is commonly observed, as in most septicemic diseases.

Symptoms and lesions may resemble other diseases so closely that a reliable diagnosis can be made only through isolation and identification of the causative organism. Good management practices that aid in preventing erysipelas include avoiding the use of ranges previously occupied by swine, sheep or turkeys where erysipelas is known to have existed. Debeaking, removal of the snoods of toms, measures that prevent injury from fighting, avoiding overcrowding and providing well drained ranges will aid in preventing this disease problem. Bacterins are available and are useful on premises where history indicates that outbreaks may be a problem. The amount and duration of protection is relative to the amount of exposure and may not be sufficient for the entire laying period. Administer bacterins in accordance with the manufacturer's directions.

Move sick birds to a hospital pen for individual treatment and to prevent cannibalism. Moving unaffected birds to a clean range may aid in preventing the spread of the disease but may also contaminate an additional range. Various antibiotics have shown efficacy in treating erysipelas; however, penicillin is best. Penicillin injections in the leg or breast muscles of visibly sick birds is effective in decreasing mortality. One injection is usually sufficient, but more may be given if necessary. Water and feed medication may be of value under certain conditions.

2

Bacteria, Growth and Reproduction

Forensic scientists often culture and grow bacteria found at crime scenes or extracted from remains. This process is often necessary to achieve a large enough population of bacteria upon which tests can then be performed.

An understanding of how bacteria grow, multiply, and change over time also helps explain many field or autopsy findings.

A population of bacteria in a liquid medium is referred to as a culture. In the laboratory, where growth conditions of temperature, light intensity, and nutrients can be made ideal for the bacteria, measurements of the number of living bacteria typically reveals four stages, or phases, of growth, with respect to time. Initially, the number of bacteria in the population is low.

Often the bacteria are also adapting to the environment. This represents the lag phase of growth. Depending on the health of the bacteria, the lag phase may be short or long. The latter occurs if the bacteria are damaged or have just been recovered from deep-freeze storage.

After the lag phase, the numbers of living bacteria rapidly increases. Typically, the increase is exponential. That is, the population keeps doubling in number at the same rate.

This is called the log or logarithmic phase of culture growth, and is the time when the bacteria are growing and dividing at their maximum speed. The explosive growth of bacteria cannot continue forever in the closed conditions of a flask of growth medium. Nutrients begin to become depleted, the amount of oxygen becomes reduced, the pH changes, and toxic waste products of metabolic activity begin to accumulate. The bacteria respond to these changes in a variety of ways to do with their structure and activity of genes. With respect to bacteria numbers, the increase in the population stops and the number of living bacteria plateaus. This plateau period is called the stationary phase. Here, the number of bacteria growing and dividing is equaled by the number of bacteria that are dying.

Finally, as conditions in the culture continue to deteriorate, the proportion of the population that is dying becomes dominant. The number of living bacteria declines sharply over time in what is called the death or decline phase. Bacteria growing as colonies on a solid growth medium also exhibit these growth phases in different regions of a colony. For example, the bacteria buried in the oldest part of the colony are often in the stationary or death phase, while the bacteria at the periphery of the colony are in the actively-dividing log phase of growth.

Culturing of bacteria is possible such that fresh growth medium can be added at rate equal to the rate at which culture is removed. The rate at which the bacteria grow is dependent on the rate of addition of the fresh medium. Bacteria can be tailored to grow relatively slow or fast and, if the set-up is carefully maintained, can be maintained for a long time. Bacterial growth requires the presence of environmental factors. For example, if a bacterium uses organic carbon for energy and structure (chemoheterotrophic bacteria), then sources of carbon are needed. Such sources include simple sugars (glucose and fructose are two examples). Nitrogen is needed to make amino acids, proteins, lipids and other components. Sulphur and phosphorus are also needed for the manufacture of bacterial components. Other elements, such as potassium, calcium, magnesium, iron, manganese, cobalt and zinc are necessary for the functioning of enzymes and other processes.

Bacterial growth is also often sensitive to temperature. Depending on the species, bacteria exhibit a usually limited range in temperatures in which they can grow and reproduce. For example, bacteria known as mesophiles prefer temperatures from 20°C–50°C (68°F–122°F). Outside this range growth and even survival is limited. Other factors, which vary depending on species, required for growth include oxygen level, pH, osmotic pressure, light, and moisture.

The events of growth and division that are apparent from measurement of the numbers of living bacteria are the manifestation of a number of molecular events. At the level of the individual bacteria, the process of growth and replication is known as binary division. Binary division occurs in stages. First, the parent bacterium grows and becomes larger. Next, the genetic material inside the bacterium uncoils from the normal helical configuration and replicates. The two copies of the genetic material migrate to either end of the bacterium. Then a cross-wall known as a septum is initiated almost precisely at the middle of the bacterium. The septum grows inward as a ring from the inner surface of the membrane. When the septum is complete, an inner wall has been formed, which divides the parent bacterium into two so-called daughter bacteria. This whole process represents the generation time.

Bacteria can exchange genetic material via conjugation. Genetic recombination between bacteria (or protists) occurs via a cytoplasmic bridge between the organisms. A primitive form of exchange of genetic material between bacteria involving plasmids also can occur. Plasmids are small, circular, extrachromosomal DNA molecules that are capable of replication and are known to be capable of transferring genes among bacteria. For example, resistance plasmids carry genes for resistance to antibiotics from one bacterium to another, while other plasmids carry genes that confer pathogenicity (the ability to cause disease). In addition, the transfer of genes via bacteriophages—viruses that specifically parasitize bacteria—also serves as a means of genetic recombination.

Bioengineering uses sophisticated techniques to purposely transfer DNA from one organism to another in order to give the second organism some new characteristic it did not have previously. For example, in a process called transformation, antibiotic susceptible bacteria that are induced to absorb manipulated plasmids placed in their environment can acquire resistance to that antibiotic substance due to the new genes they have incorporated. Similarly, in a process called transfection, specially constructed viruses are used to artificially inject bioengineered DNA into bacteria, giving infected cells some new characteristic. Evolution has driven both bacterial diversity and bacterial adaptation. Some alterations are reversible, disappearing when the particular pressure is lifted. Other alterations are maintained and can even be passed on to succeeding generations of bacteria.

Uses and Classifications of Bacteria

Bacteria are prokaryotes, that is, they are organisms that lack membranes surrounding their genetic material. Bacteria is reproduce mainly by simple, or binary, fission (splitting). Under certain condition, a mature organism splits apart to form two new organisms, it can multiply very fast and it can double its number every 20minutes. Budding is another way bacteria reproduce, it is a process in which a small out-growth from a single parent develops into a new, identical organism. There are also few bacteria that can reproduce sexually under certain conditions, this happen when two bacteria unite to form a zygote, which splits into new cells. There are some bacteria that can form a colonies big enough to be seen without using a microscope.

Uses of Bacteria

- A bacterium breaks down the organic fertilizer (decomposed vegetables and animal matter) into material that can be used by plants.
- Some species of soil bacteria convert nitrogen into nitrites, compounds that are readily absorbed by plants.
- Different commercial processes also need certain bacteria, like Anaerobic bacteria that ferment certain substances are used in the production of vinegar and some drugs, and in the aging process of cheeses.
- Some bacteria also produce waste products that are beneficial to humans, like Lactic acid, it is produced by intestinal bacteria and can promotes digestion in humans.
- Bacteria are also grown commercially and it is added to certain foods like yogurt and drinks.
- Bacteria are also used to chemically break down the tough, woody tissues of flax, jute, hemp, and coconut.
- Bacteria are also being used in modern sewage disposal known as Bioremediation, it is a process by which bacteria are added to water or soil to convert toxic pollutants, such as pesticides and oil, into harmless substances.

Bioremediation-is defined as any process that uses microorganism to return the environment altered by contaminants to its original condition. There is also a kind of bacteria that can cause somediseases, it is called pathogenic bacteria. Bacteria enter humans in different ways. Airborne bacteria enter through broken skin or by passing directly into the nose, throat, or lungs. They cause such diseases as diphtheria, pertussis (whooping cough), staph infections, strep throat, tetanus, and tuberculosis. Some bacteria enter the body means of contaminated food or and water. They can causes diseases like typhoid fever, dysentery, cholera, and botulism. Some strains of bacteria secrete toxins (poisons) into the bloodstream, which carries them to other parts of the body.

Different Uses and Classifications of Bacteria Continuation

Treatments/Prevention: Some bacterial diseases can be prevented by routine immunizations during childhood. Recommended natural remedy for bacterial infection are the following;

- Garlic-Thyme-natural and effective antibiotic. Strengthen the immune system
- Bee Propolis-another natural and strong antibiotic
- Vitamin C-a dosage of Vitamin C can help fight diseases caused by bacteria.
- Aloe Gel-loaded with amino acid and can strengthen the immune system
- Lots of water and fruit juices is important to clean the system

Classification of Bacteria based on shape:

- Bacilli (singular: Bacillus), rod-shaped bacteria; the most numerous of all types. They include coccobacilli and streptobacilli.
- Cocci-(singular: Coccus), spherical bacteria. This group is divided is divided into bacteria that occur in pairs, such as the staphylococci; in clusters, such as the streptococci.
- Spirilla-(singular: Spirillum), spiral-shaped bacteria; the least numerous type.
- Vibrios-S-shaped or comma-shaped bacteria.

Classification of Bacteria based on size, colour, and chemical composition.

- Order Actinomycetales-branching, filamentous bacteria characterized by fungus-like growth. They are important in soil fertilization and as a source of antibiotics.
- Order Beggiatoales-filamentous bacteria that move along a surface in a creeping or gliding motion. They use photosynthesis to manufacture food. Many oxidize sulphur compounds.
- Order Chlamydobacteriales-bacteria encased in a protective covering called a sheath. They grow in algae like clusters and use iron compounds for growth.

- Order Eubacteriales-rigid-cell bacteria that are rod-shaped or coccus-shaped. Most are not moving.
- Order Hyphomicrobiales-rigid cell bacteria that reproduce by budding.
- Order Myxobacteriales-flexible, rod shaped bacteria. They excrete slime and some produce appendages that are visible to the naked eye.
- Order Rickettsiales-small, rod shaped or coccus-shaped bacteria that are internal parasites in humans and other animals. Most are pathogenic.
- Order Spirochaetales-the spirochetes is a long thin, spiral shaped bacteria. Most are pathogenic.

Methods for DNA Isolation

Large Scale Double-stranded DNA Isolation

The method used for the isolation of large scale cosmid and plasmid DNA is an unpublished modification of an alkaline lysis procedure followed by equilibrium ultracentrifugation in cesium chloride-ethidium bromide gradients. Briefly, cells containing the desired plasmid or cosmid are harvested by centrifugation, incubated in a lysozyme buffer, and treated with alkaline detergent. Detergent solubilized proteins and membranes are precipitated with sodium acetate, and the lysate is cleared first by filtration of precipitate through cheesecloth and then by centrifugation. The DNA-containing supernatant is transferred to a new tube, and the plasmid or cosmid DNA is precipitated by the addition of polyethylene glycol and collected by centrifugation. The DNA pellet is resuspended in a buffer containing cesium chloride and ethidium bromide, which is loaded into polyallomer tubes and subjected to ultracentrifugation overnight. The ethidium bromide stained plasmid or cosmid DNA bands, equilibrated within the cesium chloride density gradient after ultracentrifugation, are visualized under long wave UV light and the lower band is removed with a 5 cc syringe. The intercalating ethidium bromide is separated from the DNA by loading the solution onto an equilibrated ion exchange column. The A260 containing fractions are pooled, diluted, and ethanol precipitated, and the final DNA pellet is resuspended in buffer and assayed by restriction digestion as detected on agarose gel electrophoresis.

During the course of this work several modifications to the above protocol were made. For example, initially cell growth times included three successive overnight incubations, beginning with the initial inoculation of 3 ml of antibiotic containing media with the plasmid or cosmid-containing bacterial colony, and then increasing the culture volume to 50 ml, and then to 4 l. However, it was observed that recombinant cosmid DNA isolated from cell cultures grown under these conditions, in contrast to recombinant plasmid DNA, was contaminated with deleted cosmid DNA molecules. However, these deletions are avoided by performing each of the three successive incubations for eight hours instead of overnight, although a slight yield loss accompanied the reduced growth times.

Recently, a diatomaceous earth-based method was used to isolate the plasmid or cosmid DNA from a cell lysate. The cell growth, lysis, and cleared lysate steps are performed as described above, but following DNA precipitation by polyethylene glycol, the DNA pellet is resuspended in RNase buffer and treated with RNase A and T1. Nuclease treatment is necessary to remove the RNA by digestion since RNA competes with the DNA for binding to the diatomaceous earth. After RNase treatment, the DNA containing supernatant is bound to diatomaceous earth in a chaotropic buffer of guanidine hydrochloride by incubation at room temperature. The DNA-associated diatomaceous earth then is collected by centrifugation, washed several times with ethanol buffer and acetone, dried, and then resuspended in buffer. The DNA is eluted during incubation at 65degC, and the DNA-containing supernatant is collected after centrifugation and separation of the diatomaceous earth particles. The DNA recovery is measured by taking absorbance readings at 260 nanometers. After concentration by ethanol precipitation, the DNA is assayed by restriction digestion.

Protocol:

1. Pick a colony of bacteria harbouring the plasmid or cosmid DNA of interest into a 12 X 75 mm Falcon tube containing 2 ml of LB media supplemented with the appropriate antibiotic (typically ampicillin at 100 ug/ml) and incubate at 37deg C 8-10 hours with shaking at 250 rpm. Transfer the culture to an Ehrlenmeyer flask containing 50 ml of similar media, and incubate further for 8-10 hours. Transfer 12.5 ml of the culture to each of 4 liters of similar media, and incubate for an additional 8-10 hours.
2. Harvest the cells by centrifugation at 7000 rpm for 20 minutes in 500 ml bottles in the RC5-B using the GS3 rotor. Resuspend the cell pellets in old media and transfer to two bottles, centrifuge as before, and decant the media. The cell pellets can be frozen at-70degC at this point.

3. Resuspend the cell pellets in a total of 70 ml of GET/Lysozyme solution (35 ml for each bottle) by gently teasing the pellet with a spatula and incubate for 10 minutes at room temperature. (Note: Do not vortex the lysate at any time because this may shear the chromosomal DNA).
4. Add a total of 140 ml of alkaline lysis solution (70 ml for each bottle), gently mix, and incubate for 5 minutes in an ice-water bath.
5. Add 105 ml of 3M NaOAc, pH 4.8 (52.5 ml for each bottle), cap tightly, gently mix by inverting the bottle a few times, and incubate in an ice-water bath for 30-60 minutes.
6. Clear the lysate of precipitated SDS, proteins, membranes, and chromosomal DNA by pouring through a double-layer of cheesecloth. Transfer the lysate into 250 ml centrifuge bottle, centrifuge at 10,000 rpm for 30 minutes at 4deg C in the RC5-B using the GSA rotor.

For cesium chloride-gradient purification:

7a. Pool the cleared supernatants into to a clean beaker, add one-fourth volume of 50% PEG/0.5 M NaCl, swirl to mix, and incubate in an ice-water bath for 1-2 hours.

8a. Collect the PEG-precipitated DNA by centrifugation in 250 ml bottles at 7000 rpm for 20 minutes at 4degC in the RC5-B using the GSA rotor.

9a. Dissolve the pellets in a combined total of 32 ml of 100:10 TE buffer, 5 ml of 5 mg/ml ethidium bromide, and 37 g cesium chloride (Var Lac Oid Chemical Co., Inc.) (final concentration of cesium chloride should be 1 g/ml).

10a. Transfer the sample into 35 ml polyallomer centrifuge tubes, remove air bubbles, seal with rubber stoppers, and crimp properly.

11a. Centrifuge at 60,000 rpm to 16-20 hours at 15-20degC in the Sorvall OTD-75B ultracentrifuge (DuPont) using the T-865 rotor.

12a. Visualize the ethidium bromide stained DNA under long-wave UV light, and remove the lower DNA band using a 5 cc syringe and a 25 gauge needle. (It may be helpful first to remove and discard the upper band).

13a. To remove the ethidium bromide, load the DNA sample onto an equilibrated 1.5 ml Dowex column, and collect 0.5 ml fractions. Equilibrate the Dowex AG resin (BioRad) by successive centrifugation, resuspension, and decanting with 1M NaOH, water, and then 1M Tris-HCl, pH 7.6 until the Dowex solution has a pH of 7.6.

14a. Pool fractions with an A260 of 1.00 or greater into 35 ml Corex glass tubes, add one volume of ddH2O, and ethanol precipitate by adding 2.5 volumes of cold 95% ethanol. Incubate at least 2 hours at-20degC, centrifuge at 10,000 rpm for 45 minutes in the RC5-B using the SS-34 rotor. Gently decant the supernatant, add 80% ethanol, centrifuge as before, decant, and dry the DNA pellet in a vacuum oven.

15a. Resuspend the DNA in 10:0.1 TE buffer.

For diatomaceous earth-based purification:

7b. Pool the supernatants from step 6 into 500 ml bottles and add DNase-free RNase A and RNase T1 such that the final concentration of RNase A is 40 ug/ml and RNase T1 is 40 U/ml. Incubate in a 37degC water bath for 30 minutes.

8b. Add an equal volume of isopropanol and precipitate at room temperature for 5 minutes. Centrifuge at 9,000 rpm for 30 minutes in the RC5-B using the GS3 rotor. Decant the supernatant and drain the DNA pellet.

9b. Resuspend each DNA pellet in 20 ml 10:1 TE buffer, and add 40 ml of de-fined diatomaceous earth in guanidine-HCl (100 mg/ml) to each bottle. Allow the DNA to bind at room temperature for 5 minutes with occasional mixing. Centrifuge at 9,000 for 10 minutes in the RC5-B using the GS3 rotor.

10b. Decant the supernatant, resuspend each pellet in 40 ml of diatomaceous earth-wash buffer, and centrifuge as above.

11b. Decant the supernatant, resuspend each pellet in 40 ml of acetone, and centrifuge as above.

12b. Decant the supernatant and dry the pellet in a vacuum oven.

13b. Resuspend the pellet in 20 ml of 10:1 TE buffer, and elute the bound DNA by incubation at 65degC for 10 minutes with intermittent mixing.

14b. Remove the diatomaceous earth by centrifugation at 9,000 rpm for 10 minutes in the RC5-B using the GS3 rotor. Repeat if necessary.

15b. Combine the DNA-containing supernatants and precipitate the DNA in 35 ml Corex glass tubes adding 2.5 volumes of cold 95% ethanol/acetate.

16b. Resuspend the dried DNA pellet in 2 ml of 10:0.1 TE buffer and assay for concentration by absorbance readings at 260 nm or by agarose gel electrophoresis.

Midiprep Double-stranded DNA Isolation

A midi-prep double-stranded DNA isolation has been developed to generate a sufficient amount of template DNA for several Sequenase[TM] catalyzed fluorescent terminator reactions. Here, one bacterial colony which harbored the plasmid of interest is picked into 3 ml of liquid media containing ampicillin and incubated in a 37degC shaker for 8-10 hours. At this time, the culture is transferred into 50 ml of ampicillin-containing media and incubated further for 10-12 hours.

After harvesting the cells by centrifugation, a diatomaceous earth-based alkaline-lysis purification method (19-22) is performed, similar to that discussed above for large scale DNA isolation. The purified DNA is crudely assayed for concentration and purity by agarose gel electrophoresis against known standards. The approximate yield of double-stranded DNA using this method is 1 ug of DNA per ml of cell culture. For a 50 ml cell culture, about 50 ug of DNA are recovered, and 5 ug are used typically in a Sequenase[TM] terminator reaction.

Note: This procedure is the method of choice for isolating double stranded plasmid-based templates for the Sequenase Dye-Labelled Terminator Sequencing Reactions.

Protocol:

1. Pick a colony of bacteria harbouring the plasmid DNA of interest into a 12 X 75 mm Falcon tube containing 3 ml of 2xTY media supplemented with the appropriate antibiotic (typically ampicillin at 100 ug/ml) and incubate at 37deg C 8-10 hours with shaking at 250 rpm. Transfer the culture to an Ehrlenmeyer flask containing 50 ml of similar media, and incubate further for 11-14 hours.
2. Harvest the cells by centrifugation at 3000 rpm for 5 minutes in 50 ml conical tubes in the Beckman GPR tabletop centrifuge and decant the supernatant. The cell pellets can be frozen at-70degC at this point.
3. Resuspend the cell pellets in 2 ml of GET/Lysozyme solution, add 4 ml of alkaline lysis solution, gently mix, and incubate for 5 minutes in an ice-water bath.

4. Add 4 ml of 3M NaOAc, pH 4.8, gently mix by swirling, and incubate in an ice-water bath for 30-60 minutes.
5. Clear the lysate of precipitated SDS, proteins, membranes, and chromosomal DNA by pouring through a double-layer of cheesecloth into a new 50 ml conical tube. Centrifuge at 3,000 rpm for 20 minutes at 4degC in the Beckman GPR tabletop centrifuge.
6. Decant the supernatant to a 50 ml polypropylene centrifuge tube, add 20 ul of a 20 mg/ml DNase-free RNase A and incubate in a 37degC water bath for 30 minutes.
7. Add 7 ml (equal volume) of de-fined diatomaceous earth in guanidine-HCl (20 mg/ml) and allow the DNA to bind at room temperature for 5 minutes with occasional mixing. Centrifuge at 3,000 for 5 minutes in the Beckman GPR tabletop centrifuge.
8. Decant the supernatant, resuspend in 7 ml of diatomaceous earth-wash buffer, and centrifuge as above.
9. Decant the supernatant, resuspend in 7 ml of acetone, and centrifuge as above.
10. Decant the supernatant and dry in a vacuum oven.
11. Resuspend the pellet in 0.6 ml of 10:1 TE buffer, and elute the bound DNA by incubation at 65degC for 10 minutes with intermittent mixing.
12. Remove the diatomaceous earth by centrifugation at 3,000 rpm for 5 minutes in the in the Beckman GPR tabletop centrifuge.
13. Transfer the supernatant to a 1.5 ml microcentrifuge tube and centrifuge at 12,000 rpm for 5 minutes in a microcentrifuge at room temperature. Transfer the supernatant to a new 1.5 ml microcentrifuge tube and ethanol precipitate.
14. Resuspend the dried DNA pellet in 40 ul of 10:0.1 TE buffer and assay for concentration by agarose gel electrophoresis.

Miniprep Double-stranded DNA Isolation

The standard method for the miniprep isolation of plasmid DNA includes the same general strategy as the large scale isolation. However, smaller aliquots of antibiotic containing liquid media inoculated with plasmid-containing cell colonies are incubated in a 37degC shaker for 12-16 hours. After collecting the plasmid containing cells by centrifugation, the cell pellet is resuspended in a hypotonic sucrose buffer. The cells are successively incubated with an RNase-lysis buffer, alkaline detergent, and sodium acetate.

The lysate is cleared of precipitated proteins and membranes by centrifugation, and the plasmid DNA is recovered from the supernatant by isopropanol precipitation. The DNA is crudely checked for concentration and purity using agarose gel electrophoresis against known standards. A typical yield for this method of DNA isolation is 10-15 ug of plasmid DNA from a 6 ml starting culture.

Since highly supercoiled DNA is desired for double-stranded DNA sequencing, a modification of this method employing diatomaceous earth sometimes is used for isolation of double-stranded templates for DNA sequencing with fluorescent primers. After removal of the precipitated proteins and membranes, the plasmid containing supernatant is incubated with diatomaceous earth and guanidine hydrochloride and this mixture is added into one of the twenty-four wells in the BioRad Gene Prep Manifold. The supernatant is removed by vacuum filtration over a nitrocellulose filter. The DNA-associated diatomaceous earth is washed to remove the guanidine hydrochloride with an ethanol buffer, and then dried by filtration. Elution buffer is added to the wells, and the DNA-containing solution then is separated from the diatomaceous earth particle by filtration into a collection tube. The collected DNA is concentrated by ethanol precipitation and crudely assayed for concentration and purity by agarose gel electrophoresis against known standards. The approximate yield of double-stranded DNA is 3-5 ug of DNA from 6 ml of starting culture.

Protocol:

1. Pick a colony of bacteria harbouring the plasmid DNA of interest into a 17 X 100 mm Falcon tube containing 6 ml of TB media supplemented with the appropriate antibiotic (typically ampicillin at 100 ug/ml) and incubate at 37deg C 16-18 hours with shaking at 250 rpm.
2. Harvest the cells by centrifugation at 3000 rpm for 5 minutes in the Beckman GPR tabletop centrifuge and decant the supernatant. The cell pellets can be frozen at-70degC at this point.
3. Resuspend the cell pellets in 0.2 ml of TE-RNase solution (50:10 TE buffer containing 40 ug/ml RNase A; some also add RNase T1 to a final concentration of 10 U/ul) by gentle vortexing, add 0.2 ml of alkaline lysis solution, gently mix, and incubate for 15 minutes at room temperature.
4. Add 0.2 ml of 3M NaOAc, pH 4.8, gently mix by swirling, transfer to 1.5 ml microcentrifuge tubes, and incubate in an ice-water bath for 15 minutes.

5. Clear the lysate of precipitated SDS, proteins, membranes, and chromosomal DNA by centrifugation at 12,000 rpm for 15 minutes in a microcentrifuge at 4deg C.
6. Transfer the supernatant to a fresh 1.5 ml microcentrifuge tube, incubate in an ice-water bath for 15 minutes, centrifuge as above for an additional 15 minutes and transfer the supernatant to a clean 1.5 ml tube.

For standard alkaline lysis purification:

7a. Precipitate the DNA by adding 1 ml of 95% ethanol, and resuspend the dried DNA pellet in 100-200 ul 10:0.1 TE buffer. Electrophorese an aliquot of the DNA sample on a 0.7% agarose gel to crudely determine the concentration and purity.

For diatomaceous earth-base purification:

7b. Add 1 ml of de-fined diatomaceous earth in guanidine-HCl (20 mg/ml) and allow the DNA to bind at room temperature for 5 minutes with occasional mixing. Meanwhile soak the Prep-A-Gene nitrocellulose membrane in isopropanol for at least 3 minutes, and assemble the Prep-A-Gene manifold as described in the manual.

8b. Turn on the vacuum pump and adjust the vacuum level to 8 in. Hg, let the membrane dry for 1 minute, and then release the vacuum.

9b. Pour the well mixed samples into the wells of the Prep-A-Gene manifold and filter through at 8 in. Hg until all the liquid is filtered through.

10b. Wash the samples four times with 250 ul of diatomaceous earth-wash buffer, using a repeat pipette, allowing all of the liquid to filter through between washes.

11b. Reduce the vacuum to 5 in. Hg before turning the vacuum off at the stopcock. Without unscrewing the black clamps, release the white clamps and place the collection rack with clean 1.5 ml screw-capped tubes into the manifold. Clamp the manifold with the white clamps, and apply 300 ul of 10:1 TE buffer heated to 65degC and pull the eluted DNA through at 5 in. Hg. After the liquid has filtered through, raise the vacuum to 10-12 in. Hg, and let the membrane dry for 1 minute.

12b. Turn off the vacuum at the stopcock and remove the collection rack containing the tubes. Ethanol precipitate the DNA and resuspend the dried DNA pellet in 30 ul of 10:0.1 TE buffer.

Large Scale M13RF Isolation

Double-stranded M13RF is isolated for use in M13 SmaI cut, dephosphorylated vector preparation, described below. The growth conditions of M13-infected bacterial cells appears convoluted, but result in a maximal amount of M13 RF molecules per cell. After the M13RF containing bacterial cells are harvested by centrifugation, the double-stranded molecules are isolated using the cesium chloride method for large scale plasmid isolation, as described above. This briefly entailed alkaline cell lysis, sodium acetate precipitation of detergent solubilized proteins and membranes, polyethylene glycol DNA precipitation, and extraction of ethidium bromide stained DNA from a cesium chloride gradient after ultracentrifugation.

After removal of the ethidium bromide on an ion-exchange column, the DNA containing fractions are detected by A260 measurement and pooled, and the DNA is concentrated by ethanol precipitation and assayed by restriction enzyme digestion and agarose gel electrophoresis.

Protocol:

1. Prepare an early log phase culture of JM101 by inoculating an Ehrlenmeyer flask containing 50 ml of 2xTY with a glycerol stock of JM101 and pre-incubating for 1 hour in a 37degC water bath, with no shaking. Pick a plaque representing the desired M13 clone into four 1.5 ml aliquot of early log phase JM101, and incubate according to the procedure displayed in Figure 1 to result in 4 liters of M13-infected bacteria.
2. Harvest the cells by centrifugation at 7000 rpm for 20 minutes in 500 ml bottles in the RC5-B using the GS3 rotor. Resuspend the cell pellets in fresh 2xTY media to remove contaminating extracellular phage and transfer to two bottles, centrifuge as before, and decant the media. The cell pellets can be frozen at-70degC at this point.
3. Resuspend the cell pellets in a total of 120 ml (30 ml for each bottle) of 1X STB buffer by gently teasing the pellet with a spatula. Add a total 24 ml of lysozyme solution (6 ml for each bottle), gently mix, and incubate for 5 minutes in an ice-water bath.
4. Add 48 ml of 50:2:10 TTE buffer (12 ml for each bottle) and 2 ml of RNase A (10 mg/ml) (0.5 ml for each bottle), gently mix, and incubate in an ice-water bath for 5 minutes.
5. Clear the lysate of precipitated SDS, proteins, membranes, and chromosomal DNA by pouring through a double-layer of cheesecloth. Transfer the lysate into 250 ml centrifuge

bottle, centrifuge at 10,000 rpm for 30 minutes at 4deg C in the RC5-B using the GSA rotor.

6. Add 6 ml of 5 mg/ml ethidium bromide, and cesium chloride such that the final concentration of cesium chloride is 1 g/ml.
7. Transfer the sample into 35 ml polyallomer centrifuge tubes and top off with a 1:1 solution of 100:10 TE buffer and cesium chloride, remove air bubbles, seal with rubber stoppers, and crimp properly.
8. Centrifuge at 60,000 rpm to 16-20 hours at 15-20degC in the Sorvall OTD-75B ultracentrifuge using the T-865 rotor.
9. Visualize the ethidium bromide stained DNA under long-wave UV light, and remove the lower DNA band using a 5 cc syringe and a 25 gauge needle. (It may be helpful to remove and discard the upper band first).
10. To remove the ethidium bromide, load the DNA sample onto an 1.5 ml Dowex AG (BioRad) column, equilibrated as before, and collect 0.5 ml fractions.
11. Pool fractions with an A260 of 1.00 or greater into 35 ml Corex glass tubes, add one volume of ddH2O, and ethanol precipitate by adding 2.5 volumes of cold 95% ethanol. Incubate at least 2 hours at-20degC, centrifuge at 10,000 rpm for 45 minutes in the RC5-B using the SS-34 rotor. Gently decant the supernatant, add 80% ethanol, centrifuge as before, decant, and dry the DNA pellet in a vacuum oven.
12. Resuspend the DNA in 10:0.1 TE buffer.

Single-stranded M13 DNA Isolation Using Phenol

This isolation procedure is the method of choice for preparation of M13-based templates to be used in Sequenase[TM] catalyzed dye-terminator reactions. A pre-incubated early log phase JM101 culture is prepared by transferring a thawed glycerol stock into 50 ml of liquid media and incubating for 1 hour at 37degC with no shaking.

M13 plaques are picked with a sterile toothpick and placed into 1.5 ml aliquots of the early log phase JM101 culture, which are incubated in a 37deg C shaker for 4-6 hours. After incubation, the bacterial cells are pelleted by centrifugation and the viral containing supernatant is transferred to a clean tube.

The phage particle are precipitated with PEG, collected by centrifugation, and the pellet is resuspended in buffer. The phage protein coat is denatured and removed by one phenol and two ether extractions. After ethanol precipitation, the dried DNA pellet is resuspended in buffer, and the concentration and purity crudely are assessed by agarose gel electrophoresis against known standards.

Protocol:

1. Prepare an early log phase culture of JM101, as above, and pick M13-based plaques with sterile toothpicks into 12 X 75 mm Falcon tubes containing 1.5 ml aliquots of the cells. Incubate for 4-6 hours at 37degC with shaking at 250 rpm.
2. Transfer the culture to 1.5 ml microcentrifuge tubes and centrifuge for 15 minutes at 12,000 rpm at 4degC.
3. Pipette the top 1 ml of supernatant to a fresh 1.5 ml microcentrifuge tube containing 0.2 ml 20% PEG/2.5 M NaCl to precipitate the phage particles. Mix by inverting several times and incubate for 15-30 minutes at room temperature.
4. Centrifuge for 15 minutes at 12,000 rpm at 4degC to collect the precipitated phage. Decant the supernatant and remove residual PEG supernatant by suctioning twice.
5. Resuspend the pellet in 100 ul of 10 mM Tris-HCl, pH 7.6 by vortexing, and add 50 ul of TE-saturated phenol.
6. Extract the DNA with phenol and twice with ether, as discussed above, and then ethanol precipitate.
7. Resuspend the dried DNA in 6 ul of 10:0.1 TE for use in single-stranded Sequenase[TM] catalyzed dye-terminator sequencing reactions.

Biomek-automated Modified-eperon Isolation Procedure for Single-stranded M13 DNA

This semi-automated method is a modification of a previously reported procedure, and allowed the simultaneous isolation of 48 single-stranded DNAs per Biomek 1000 robotic workstation within 3 hours. Basically, M13 plaques are picked with sterile toothpicks into aliquots of early log phase JM101, prepared as discussed above. The phage infected cultures are incubated in a 37degC shaker for 4-6 hours, transferred into microcentrifuge tubes, centrifuged to separate bacterial cells from the viral supernatant, and then carefully placed on the Biomek tablet. For each sample, two 250 ul aliquots are robotically distributed into two wells of a 96-well microtiter plate, and this process is repeated for each of the 48 samples until the entire 96 wells are filled. A solution of polyethylene glycol (PEG) then is added robotically to each well and mixed. The microtiter plate is covered with an acetate plate sealer, incubated at room temperature to precipitated the phage particles, and then centrifuged. The supernatant then is removed by inverting the plate and gently draining on a paper towel, without dislodging the pellet. After placing the microtiter plate back on the Biomek, a more dilute PEG solution is robotically added to each well. The plate then is covered with another sealer and centrifuged again. This rinse step aided in the removal of contaminating proteins and RNA. After removing the supernatant, as before, and placing the microtiter plate back on the Biomek, a Triton X-100 detergent solution is robotically added to each well. The plate is agitated gently and the sample from each pair of wells is robotically transferred to microcentrifuge tubes, which then are capped and placed in an 80deg C water bath for 10 minutes to aid in the detergent solubilization of phage coat proteins. After a brief centrifugation to collect condensation, the single-stranded DNA is ethanol precipitated, dried, and resuspended. An aliquot from each DNA sample is subjected to agarose gel electrophoresis to crudely assay concentration and purity. The yield of single-stranded template is approximately 2-3 ug per sample.

Protocol: The entire procedure will require 9 rows of P250 tips (counting from the centre of the Biomek tablet towards the left) for the isolation of 48 templates (48ISOL). The reagent module should contain PEG-2000, Triton-Tris-EDTA, and ethanol-acetate, respectively.

1. Prepare an early log phase JM101 culture in 50 ml of 2xTY, as above.
2. Using sterile toothpicks, transfer individual M13 plaques into 12 X 75 mm Falcon tubes containing 1 ml early log phase cell cultures, and incubate for 4-6 hours at 37degC with shaking at 250 rpm. (Growth for longer than 6 hours results in cell lysis and contamination of the phage DNA by cellular proteins and nucleic acids).
3. Separate the bacterial cells from the viral-containing supernatant by centrifugation at 12,000 rpm for 15 minutes at 4degC. Carefully open the tubes and place on the Biomek tablet.
4. The Biomek will distribute two 250 ul aliquots of viral supernatant per sample into the wells of a 96-well flat-bottomed microtiter plate (Dynatech). The Biomek then will add 50 ul of 20% PEG/2.5 M NaCl solution to each well, and mix by pipetting up and down.

5. Cover the plate with an acetate plate sealer and incubate at room temperature for 15 minutes.
6. Pellet the precipitated phage by centrifuging the plate at 2400 rpm for 20 minutes in a Beckman GPR tabletop centrifuge. Remove the plate sealer and drain the PEG from the plate by gently draining upside down on a Kimwipe.
7. Return the plate to the tablet, and the Biomek will robotically add 200 ul of PEG:TE rinse solution to each well. Cover the plate with a plate sealer, centrifuge, and drain, as above.
8. Return the plate to the tablet, and the Biomek will add 70 ul of TTE solution to each well. Remove and gently agitate to resuspend.
9. The Biomek then will robotically pool the contents from each pair of wells into 1.5 ml microcentrifuge tubes.
10. Incubate the tubes at 80degC for 10 minutes to denature the viral protein coat and then centrifuge briefly to reclaim condensation.
11. Ethanol precipitate the DNA by adding 500 ul ethanol/acetate to each tube, as described above.
12. Resuspend the DNA templates in 20 ul of 10:0.1 TE buffer.

96 Well Double-stranded Template Isolation

A manual as well as an automated procedure is given below. The automated method is a modification of a previously reported procedure which allows simultaneous isolation of 96 double stranded DNAs per Biomek 1000 Automated Laboratory Workstation within two hours. Basically colonies containing double-stranded plasmids are picked with sterile toothpicks into media and incubated at 37degC for 24 hours with shaking at 350 rpm. These cells are harvested by centrifugation and the pellets are either manually or robotically resuspended by the addition of TE-RNase solution. An alkaline lysis solution is used to lyse the cells and the lysate is precipitated with KOAc. The lysate is cleared by filtration and further concentrated by ethanol precipitation. An aliquot from each DNA sample is subjected to agarose gel eletrophoresis to crudely assay concentration and purity. The yield of double stranded template is approximately 3 mg per sample.

Protocol: Manual Double stranded isolation method: The following is a manual, 96 well, double stranded sequencing template isolation procedure that has been developed in our laboratory. A similar procedure that has been automated on the Biomek is presented elsewhere herein.

1. Pick individual shotgun clones off a plate with a steril tooth pick and deposit each separately into 96 well block containing 1.75 ml of TB media per well. Keep toothpick in media for about 5 minutes to allow the cells to defuse into the media, remove the toothpicks, cover the 96 well block with the loose fitting lid, and allow the cells to grow for 24 hours in the 37degC shaker/incubator at 350 rpm.
2. Remove block from the shaker/incubator and collect the cells by centrifugation at 2500 rpm for 7 minutes. The cells can be stored frozen at -20degC in the block at this stage.
3. After thawing the cells, add 100ul TE-RNAse-A solution containing RNAse T1, mix by pipetting up and down 4-5 times to resuspend the cell pellet and then incubate in the 37degC incubator/shaker for 5 minutes at 350 rpm to mix more thoroughly.
4. Remove the block from the incubator/shaker and then add 100ul of alkaline lysis solution. Shake the block by hand to mix the reagents and then incubate at room temperature for 1 hour with intermittent swirling. 5. Then add 100ul of either 3M potassium or sodium acetate, pH 5, and place the block in the 37degC shaker/incubator for 5 minutes at 350 rpm to thoroughly mix and shear genomic DNA to reduce the viscosity of the solution. Place the block at -20degC for 30 minutes.
6. Centrifuge the block in the GPR centrifuge at 3000 rpm at 4degC for 30 minutes.
7. Carefully remove 200 ul of the supernatant from each well in the 96 well block with the 12 channel pipetter and transfer them to a v-bottom microtiter plate, being careful not to transfer any cell debris.
8. Transfer 10 ul of supernatant into the respective cycle sequencing reaction tubes, and precipitate with 150 ul of 95% ethanol (without added acetate). After storage at -20degC for 30 minutes, the pellet is collected by centrifugation, washed three times with 70% ethanol, and dried directly in the cycle sequencing reaction tubes.
9. Prior to adding the fluorscent terminator cycle sequencing reaction mix, the dried templates should be stored at -20degC. An additional 75 ul of the supernatant is transferred to a Robbins PCR reaction tube (in 96 well tube format) and precipitated with 200 ul of 95% ethanol, washed three times with 70% ethanol, and stored dry at -20degC for future use.

The following is an automated, 96 well, double stranded sequencing template isolation procedure that has been developed in our laboratory.

1. Pick colonies using a toothpick into 1.8 ml TB with TB salts containing appropriate antibiotic and shake for 22-24 hours at 350 rpm in a 96 well block with cover.
2. Harvest cells by centrifugation at 1800 rpm for 7 min. Pour off supernatant and allow pellets to drain inverted. Cell pellets may be frozen at this point if necessary.
3. Turn on Biomek, begin the program DSISOL2 and set up the Biomek as indicated in the configuration function on the screen. Specifically, you should put TE-RNase solution in the first module, alkaline lysis solution in the second reagent module and 3 M KOAc, pH 4.8 in the third module.
4. Place the 96 well block containing cells onto the Biomek tablet at the position labelled "1.0 ml Minitubes". Place a Millipore filter plate in the position labelled "96well flat bottomed microtitre plate".
5. Press ENTER to continue with the program.
6. First the Biomek will add 100 ml TE-RNase solution to the cell pellets and mix to partially resuspend.
7. Next, the biomek will add 100 ml alkaline lysis solution to the wells of the filter plate.
8. The biomek then will mix the cell suspension again, transfer the entire volume to the filter plate containing alkaline lysis solution, and mix again. Set up the filtration apparatus with a clean 96 well block to collect the filtrate (wash and reuse the block used for growth).
9. The biomek will add 100 ml 3M KOAc, pH 4.8 to the wells of the filter plate and mix at the sides of the wells. Some choose to place the filter plate at -20degC for 5 minutes at this point. Transfer the filter plate to the QiaVac Vacuum Manifold 96 and filter using water vacuum only (do not do a harsh filtration as the plates are fragile and will loose their seal). This will typically take less than 20 minutes.
10. The supernatant collected in the 96 well block is the crude DNA and must be ethanol precipitated before use by the addition of 1 ml 100% ethanol and incubation at -20degC for at least 30 minutes.
11. Centrifuge for 25 minutes at 3000 rpm in a cooled Beckman GPR centrifuge.

12. Decant and wash with 500 ml 80% ethanol and centrifuge for an additional 5 minutes at 3000 rpm.
13. Decant the supernatant, drain inverted on a paper towel. Dry under vacuum.
14. Resuspend in 50 ml 10:0.1 TE for use in dye primer or dye terminator sequencing chemistry.

Genomic DNA Isolation from Blood

Genomic DNA isolation is performed according to the FBI protocol. After the blood samples (stored at -70degC in EDTA vacutainer tubes) are thawed, standard citrate buffer is added, mixed, and the tubes are centrifuged. The top portion of the supernatant is discarded and additional buffer is added, mixed, and again the tube is centrifuged. After the supernatant is discarded, the pellet is resuspended in a solution of SDS detergent and proteinase K, and the mixture is incubated at 55deg C for one hour. The sample then is phenol extracted once with a phenol/chloroform/isoamyl alcohol solution, and after centrifugation the aqueous layer is removed to a fresh microcentrifuge tube. The DNA is ethanol precipitated, resuspended in buffer, and then ethanol precipitated a second time. Once the pellet is dried, buffer is added and the DNA is resuspended by incubation at 55degC overnight, the genomic DNA solution is assayed by the polymerase chain reaction.

Protocol:

1. Blood samples typically were obtained as 1 ml of whole blood stored in EDTA vacutainer tubes frozen at -70deg C.
2. Thaw the frozen samples, and to each 1 ml sample, add 0.8 ml 1X SSC buffer, and mix. Centrifuge for 1 minute at 12,000 rpm in a microcentrifuge.
3. Remove 1 ml of the supernatant and discard into disinfectant.
4. Add 1 ml of 1X SSC buffer, vortex, centrifuge as above for 1 minute, and remove all of the supernatant.
5. Add 375 ul of 0.2M NaOAc to each pellet and vortex briefly. Then add 25 ul of 10% SDS and 5 ul of proteinase K (20 mg/ml H2O) (Sigma P-0390), vortex briefly and incubate for 1 hour at 55degC.
6. Add 120 ul phenol/chloroform/isoamyl alcohol and vortex for 30 seconds. Centrifuge the sample for 2 minutes at 12,000 rpm in a microcentrifuge tube.
7. Carefully remove the aqueous layer to a new 1.5 ml microcentrifuge tube, add 1 ml of cold 100% ethanol, mix, and incubate for 15 minutes at -20deg C.

8. Centrifuge for 2 minutes at 12,000 rpm in a microcentrifuge. Decant the supernatant and drain.
9. Add 180 ul 10:1 TE buffer, vortex, and incubate at 55degC for 10 minutes.
10. Add 20 ul 2 M sodium acetate and mix. Add 500 ul of cold 100% ethanol, mix, and centrifuge for 1 minute at 12,000 rpm in a microcentrifuge.
11. Decant the supernatant and rinse the pellet with 1 ml of 80% ethanol. Centrifuge for 1 minute at 12,000 rpm in a microcentrifuge.
12. Decant the supernatant, and dry the pellet in a Speedy-Vac for 10 minutes (or until dry).
13. Resuspend the pellet by adding 200 ul of 10:1 TE buffer. Incubate overnight at 55degC, vortexing periodically to dissolve the genomic DNA. Store the samples at -20degC.

Plasmids

A plasmid is an independent, circular, self-replicating DNA molecule that carries only a few genes. The number of plasmids in a cell generally remains constant from generation to generation. Plasmids are autonomous molecules and exist in cells as extrachromosomal genomes, although some plasmids can be inserted into a bacterial chromosome, where they become a permanent part of the bacterial genome. It is here that they provide great functionality in molecular science.

Plasmids are easy to manipulate and isolate using bacteria. They can be integrated into mammalian genomes, thereby conferring to mammalian cells whatever genetic functionality they carry. Thus, this gives you the ability to introduce genes into a given organism by using bacteria to amplify the hybrid genes that are created in vitro. This tiny but mighty plasmid molecule is the basis of recombinant DNA technology.

There are two categories of plasmids. Stringent plasmids replicate only when the chromosome replicates. This is good if you are working with a protein that is lethal to the cell. Relaxed plasmids replicate on their own. This gives you a higher ratio of plasmids to chromosome.

So how do we manipulate these plasmids?

1. Mutate them using restriction enzymes, ligation enzymes, and PCR. Mutagenesis is easily accomplished by using restriction enzymes to cut out portions of one genome and insert them into a plasmid. PCR can also be used to facilitate mutagenesis.

Plasmids are mapped out indicating the locations of their origins of replication and restriction enzyme sites.

2. Select them using genetic markers. Some bacteria are antibiotic resistant. While this is a serious health problem, it is a godsend to molecular scientists. The gene that confers antibiotic resistance can be added (ligated) to the gene you are inserting into the plasmid. So every plasmid that contains your target gene will not be killed by antibiotics. After you transfect your bacterial cells with your engineered plasmid (the one with the target gene and the antibiotic resistant marker), you incubate them in a nutrient broth that also contains antibiotic (usually ampecillin). Any cells that were not transfected (this means they do not have your target gene in them) are killed by the antibiotic. The ones that do have the gene also have the antibiotic resistant gene, and therefore survive the selection process.
3. Isolate them (such as with alkaline lysis)
4. Transform them into cells where they become vectors to transport foreign genes into a recipient organism.

There are some minimum requirements for plasmids that are useful for recombination techniques:

1. Origin of replication (ORI). They must be able to replicate themselves or they are of no practical use as a vector.
2. Selectable marker. They must have a marker so you can select for cells that have your plasmids.
3. Restriction enzyme sites in non-essential regions. You don't want to be cutting your plasmid in necessary regions such as the ORI.

In addition to these necessary requirements, there are some factors that make plasmids either more useful or easier to work with.

1. Small. If they are small, they are easier to isolate (you get more), handle (less shearing), and transform.
2. Multiple restriction enzyme sites. More sites give you greater flexibility in cloning, perhaps even allowing for directional cloning.
3. Multiple ORIs. It is important to note that two genes must have different ORIs if they are going to be inserted in the same plasmid.

R-Plasmids

R-Plasmids-It was long known that repeated exposure of bacteria to antibiotics results in the development of a resistant strain. Some bacterial cells undergo gene mutation which makes them resistant to the antibiotic. They live and multiply 10 the presence of the drug, giving rise to a resistant strain. This type of resistance develops gradually, and can be treated by administering antibiotics before a large number of resistant cells are formed.

Another type of resistance was observed with respect to the antibiotic gentamycin. Bacterial strains developed resistance to gentamycin only, a few months after its introduction. The gentamycin-resistant strain was to also found to be resistant to three other antibacterial agents, chloramphenicol, ampicillin and sulfonamide Growth of the strain with any susceptible strain of bacteria conferred antibiotic resistance to the latter within a few hours. Thus the antibiotic resistance was infectious.

The infectious antibiotic resistance factor or R factor consists of genes producing resistance to the four antibiotics and genes that facilitate the, transfer of the R factor. They are carried in plasmids called the R (for resistance) plasmids. When a strain carrying R plasmids conjugates with other bacterial cells, a copy of the R plasmid is transferred to the latter, giving it the property of drug resistance.

R plasmids were first discovered in a strain of Shigella in the late 1950s during an epidemic of dysentery in Japan. Shigella acquired drug resistance with the introduction of antibiotic therapy in 1946. By 1964, 50% of shigellas were resistant to all the four major drugs (sulfonamides, streptomycin, tetracycline and chloramphenicol) against dysentery. Plasmid-mediated drug resistance has serious medical impli-cations. Firstly contact with a drug-resistant non-pathogenic member of the normal gut flora can make a drug-sensitive pathogen resistant. Secondly, a pathogen may acquire new R plasmid resistances during epidemics. Plasmid DNA occurs in two forms called co-integrates and aggregates. The plasmid co-integrate contains all the resistance genes, together with the genes required for transfer, in one circular molecule, e. g. RIGO (Cm, Sm Tu, Tc) isolated from Shigella flexneri.

In plasmid aggregates the various genes are located, at least for some time, on separate molecules, e. g. in one isolate of Salmonella typhimurium type 29.

3

Plant Protection

The State Plant Protection Department (hereinafter referred to as Department) is guiding the policy of state plant protection. In order to protect the territory of Lithuanian Republic from harmful organisms, diseases as well as from import and distribution of weeds, the Department performs the inspection of imported plants and researches of plant diseases, pests and weeds pervasion.

The Department is responsible for the certification of plant protection means as well as their usage control.

The State Plant Protection Department prepares projects in the field of plant protection and quarantine and adjust them according to the EU standards and other international agreements.

The State Plant Protection Department consists of 6 Administration Divisions, 10 Regional Plant Protection and Quarantine Centres (these centres conduct researches of harmful organisms pervasion in agricultural sown areas and participate in creating forecasts of agricultural pests occurrence as well as perform control of imported, exported plants and vegetable products.

These centres are also responsible for issuance of phytosanitary certificates and provision of consultations for users of plant protection means) as well as 19 Borderline Plant Quarantine Centres (these centres are responsible for borderline control of imported, exported plants and vegetable products).

The Department employs 250 workers, about 50 employees work in the Administration, in Vilnius.

Preparation for the EU Membership

When executing plans of Law Coordination Means and Acquis Implementation Means as well as carrying out the Negotiation position covenants, all necessary acts were prepared and ratified in 2003 that were necessary for adjustment of Lithuanian Republic Phytosanitary law base according to the legal acts of the EU. During 2003, legal base of the EU was systematized; information about new legal acts of the EU in the field of phytosanitary and plant protection was collected and analysed. Tables of conformity to the EU directives were prepared for all national phytosanitary legal acts adjusted according to the EU requirements. These tables indicated the conformity of legal acts and their compatibleness with the requirements of the Union. A meeting with the officials of European Committee was summoned in order to create conformity tables of national legal acts to the EU legal acts. The experts of European Committee analysed the tables and commended them.

The order of plant protection products certification, usage, storage and trade is established in Lithuanian Republic Law on Plant Protection that shall become valid from 1 May 2004. Conformity to the requirements of mentioned Law shall be of great importance for farmers, because their allowances will depend on the proper usage of plant protection products. The European Union promotes the preparation of new member states for the membership and provides a financial support, i.e. PHARE program and bilateral partnership projects.

Quality Control and Analysis of Diseases

According to the quality requirements, in 2003 inspectors examined more than 140 000 t of imported fresh fruit and vegetables originating from 34 different countries. In 2003, Holland (onions, carrots and salad vegetables), Spain (cucumbers, tomatoes, exotic fruits), Poland (apples) were the main suppliers. More than 93 percent of this production conformed to the mandatory quality requirements. The main reasons of nonconformity were external quality defects (about 540 cases), wrong labelling or labelling absence (about 130 cases), weight (7 cases). In order to eliminate the mentioned incompliance, the retailers of fresh fruit and vegetables usually employed the method of quality class reduction (this method was applied for 2936 t) and re-assortment of products (1117 t were re-sorted) so that products would conform to the mandatory quality requirements. The similar situation was in 2002 m.

In 2003, production of Lithuanian farmers was subjected to more frequent inspections. More kinds of fruit and vegetables were inspected than in 2002. 114 cases out of 296 inspections performed showed that Lithuanian production did not conform to the quality requirements. In 2003, quality inspectors registered cases of nonconformity. The number of such cases was by 30 percent lower than in 2002. Owners of companies should not be considered as the only persons that ensured better preparation of production for sale. The Department experts that conducted seminars and provided with other information about mandatory quality requirements had also significantly contributed to it. In 2003, 2437 storage, trade and usage areas of plant protection products were inspected and as a result there were 1550 cases of major or minor violations registered.

In 2003, 2331 certificates of conformity to the quality requirements were issued.

Last year, 8478 inspections of accountancy and phonological examinations were performed. The information on researched diseases and pests is presented in the descriptions.

According to the consumed amount of plant protection products during 2003, it was established that an average of 0.25 kg of active agents was used for one hectare of agricultural landed property. In 2003, 1681.8 thousand ha of plants were subjected to the plant protection products. In 2003, 938 000 ha were subjected to herbicides in order to fight weeds (this number is by 78.9 thousand ha more than in 2002, and 419,6 thousand ha more than 1995).

Status of Protected Zones

According to the agreement of joining into the EU, Territory of Lithuanian Republic is declared as a protected area from fruit-tree bacterial blight (Erwinia amylovora) (hereinafter referred to as blight) and Beet Necrotic Yellow Vein Virus (Rhizomania) (hereinafter Rhizomania) that are spread in other EU countries as well as spread in specialised cultures that have a limited economic importance to the entire EU.

Certain plants that may carry such pests or diseases must have plant passport with a special code, consisting of letters “ZP” (i.e. zone protected), when imported into Lithuania or other protected zones. This passport must also contain codes of all countries that have a protected zone status in the EU. These codes indicate that plants conform to the EU requirements for certain pests or diseases and can be imported into the certain protected zone.

Plant Passport

From 1 January 2004, according to the ratified legal acts, the new labelling system that conforms to the phytosanitary and quality requirements of the EU became in operation, i.e. the issuance of plant passports was introduced. Plant passport is a document valid within the entire EU; it is intended to ensure conformity of plants, vegetable production and etc. to the phytosanitary requirements. So from now on, Lithuanian vegetable production shall have equal opportunities to be realized not only in the local market, but also in the EU countries.

In 2003, 84 611 phytosanitary certificates and 1022 phytosanitary register certificates were issued; 2292 phytosanitary inspections of plants and vegetable production were performed.

Performance Improvement

The employees of the Department intensively raise their qualification as well as gather information on the news in their speciality. In 2003, they participated in 20 International seminars and traineeships in the field of plant healthiness, introduction of plant passport system, bacteriological blights and Rhizomania. Seminars were summoned in the Department as well. These seminars were conducted by foreign experts that visited Lithuania 20 times last year. In 2003, the Department continued execution of the Twins project „Modernisation of phytosanitary researches and administration", dated 2001. Six foreign experts provided with consultations in the field of wood, potato growing, plant passport and protected zones. In 2003, the PHARE project „Consolidation of General Agricultural Policy" was started. One part of this project is intended for development of quality control system in Lithuania, German expert provided consultations about the issue.

The Department organized about 30 courses and seminars on various issues of quality control for inspectors working in Regional Plant Protection and Quarantine Centres as well as in Borderline Plant Quarantine Centres. The Department and Lithuanian Horticulture and Olericulture Institute trained 21 inspectors on the performance of quality checks according to the 34 fruit and vegetable positions of Composite Product nomenclature valid in the EU.

5512 users were trained by using the Curriculum for Plant Protection Means Users approved by Ministry of Agriculture, certificates of plant protection courses completion were issued that grant a right to purchase and use products intended for professional usage. By using learning program for sellers of plant protection products, 157 courses were conducted and 374 persons, responsible for trade and storage of plant protection products, were trained. During 2003, 5886 certificates of completion of plant protection courses were issued. In 2003, about 300 seminars and courses for agriculturalists were held that provided information about requirements for storage, usage and accountancy of plant protection products as well as about product control system and responsibility for violations. There were more than 60 publications on the above issues in the national press.

Electronic system Caddy for registration of applications for plant protection products was introduced. This new system will help to reduce costs and time for processing of applications.

Lithuanian Borderline

In 2003, two last Borderline Phytosanitary Control Posts in Medininkai and Lavoriškes were finished and equipped with the necessary equipment conforming to the EU requirements. The building of these posts was financed by the PHARE project (2000) and national budget. In the beginning of 2003, the quality inspectors, working in the Borderline and Regional Plant Protection and Quarantine Centres were provided with the necessary equipment for analysis of fresh fruits and vegetables. According to the legal acts, as from 1 January 2004, three kinds of Ambrosia spp. were included into the harmful parasitic plant lists. From now on, the imported grain into Lithuania shall be controlled depending from the origination country (i.e. from EU countries or not EU countries) in order to protect Lithuanian territory from importing and spreading of weeds.

Ground Application Equipment

Anyone with access only to the published records of ground spraying against tsetse would have difficulty in discovering just how the insecticides are applied. This is being remedied to some extent by the publication of more information on pesticide application equipment, a good deal of which has relevance to tsetse control. COPR and the Overseas Spraying Machinery Centre (OSMC) in collaboration with other UK scientific units through the Ministry of Overseas Development Natural Resources Advisors Coordinating Committee (ODM NRACC) Sub-Committee on Pesticide Application Overseas help in the promotion and coordination of pesticide application research and development in relation to the demands of overseas countries. Assessment of spraying machinery is ongoing research at OSMC, and many of the tests carried out are based upon recommendations by WHO. Spraying equipment is used to disseminate pesticides as liquids in the form of aqueous suspensions, emulsions and solutions. It is not possible to list all the equipment in current use or which is available, but the apparatus mentioned below has been used in major control operations; e.g. aerosol dispensers are used at fly pickets, fogging machines, although employed in the early days, have now declined in use, whereas a variety of hydraulic energy and gaseous energy equipment is still in regular service.

Hydraulic Energy Equipment

Spray Nozzles

There are three basic pattern shapes produced by hydraulic energy nozzles; the falt fan, the solid cone and the hollow cone. The limitations of the fan pattern for tsetse work are reflected in the very few occasions on which it appears to have been used. In a campaign against G. palpalis, the Eclipse sprayer was used, fitted with a lance that gave a fan-shaped spray for treating the leaves and stems of plants along paths. One difficulty which arises with these nozzles is that the width of the fan has to be orientated in order to cover the surface to be sprayed.

Cone-shaped sprays are produced by nozzles which consist essentially of a swirl plate with angled slits or holes which produce a vortex in the flow of insecticide, causing it to issue from the orifice in the form of a solid or hollow cone of spray depending on the presence or absence of a central hole in the core. Whether the solid cone nozzle has ever been used in tsetse control is uncertain.

The hollow cone spray nozzle is in general use and either a constant or variable spray width can be produced. For the former, the actual dimensions are determined by the design of the swirl core, the size of the nozzle disc orifice and the operating pressure. The practical limitations of these nozzles is that when spraying tree trunks or individual branches, the position of the nozzle, in relation to the surface being sprayed, must be constantly adjusted to ensure that a swath of insecticide of the required width is obtained.

Accurate and, therefore, less wasteful application can be achieved by a skilled operator using a spray lance (sometimes referred to as "gun") emitting a cone of spray whose shape can be instantly adjusted to fit the width of the object being treated. This can be achieved by altering the length of the swirl chamber. The angular width of the cone at a given pressure is inversely proportional to the length of the chamber between the swirl block and orifice disc. Alternatively, wastage can be avoided by adjusting the distance between the nozzle and the target surface, but this may necessitate moving the nozzle more rapidly when close to the surface.

Spray Lances

Two types of spray lances are in general use; one has a shut-off valve in the handle for controlling the flow of insecticide and produces a constant or variable cone depending upon the type of nozzle fitted. With the other, the shape of the cone is adjusted from the handle by means of a control rod inside the lance.

Shut-Off Valve Models

These consist of a trigger-operated, instantaneous shut-off valve which, with tube and nozzle, from the lance assembly. To produce a constant-shape cone, the nozzle assembly contains a swirl cone, an orifice disc and a screw-retaining cap. The shape of the cone produced for a given pressure is related to both the design of the swirl plate and the size of the orifice in the disc. Variable cone nozzles consist of a nozzle body attached to the end of the lance by means of an adaptor. Variations in the shape of the cone are produced by altering the length of the swirl chamber between the core and the disc by screwing or unscrewing the cap. Leakage of insecticide down the lance from the nozzle is prevented by an 'O' ring immediately above the threaded portion of the body.

The CP 601 spray gun has been used in some countries. This consists of a trigger-operated shut-off valve with a brass body, a plated steel trigger, a brass tube 58.5 cm in length with a 45° bend near the end and a single cone spray nozzle No. 520/P. The nozzle has a plastic swirl core, of which only one model is available, a stainless steel nozzle disc and a plastic nozzle cap. Six different orifice sizes are available, with diameters ranging from 0.79 to 2.38 mm, but in both Nigeria and Uganda disc No. 12 with a 1.19 mm orifice is fitted.

Also in this category is the Stoprex spray gun. This has a thumb trigger in contrast to the models previously described, all of which have a four-finger trigger for operating the shut-off valve. The lance and constant-shape cone nozzle assembly are, however, similar to those of the CP601, but the Rex Variable Nozzle is of a different design. Variations in swirl chamber length are made by rotating a collar to which the same stainless steel orifice disc and plastic nozzle cap as those used for hollow cones are fitted.

Control Rod Models

Spray guns of this type are intended for use in situations where frequent and often rapid changes in the width of the spray pattern are necessary, e.g. when treating the trunks of different trees and the underside of branches at varying distances from the operator. As with the variable cone nozzles described above, pattern shape is changed by altering the length of the swirl chamber. The disc is fixed but movement of the core is achieved by the control rod. The CP700 Spray Gun is one example of this model which has been extensively used in Uganda but is no longer available. A some what different mechanism for adjusting cone shape is found in the GunJet No. 12 and No. 14 spray lances whose use in Botswana has been reported. The essential difference between these and the No. 700 is that changes in the swirl chamber are made by rotating the control rod. On rotating the handle the swirl chamber is progressively opened and spray is emitted, at first in the form of a short broad cone.

As this movement is continued, the cone becomes progressively longer and narrower until a complete circle has been turned and the swirl chamber is fully open and the spray is emitted as a solid stream. The cone dimensions are related to operating pressure and orifice size and two types of fitting may be used for producing hollow cones. One is a stainless steel disc intended for the higher rates of application, while for the lower outputs the disc can be replaced by a ConeJet spray tip, of which ten sizes are available.

Another lance in which the spray cone is adjusted by means of a trigger-operated control rod has been used with a Platz vehicle-mounted sprayer in Zambia. The output of each lance is controlled by the manual operation of the trigger and can be varied between a very fine cone spray and a solid jet (37). The maximum output (from one lance) is about 4.1 per minute with a 2 mm disc orifice and 2.18 per minute with a 1.5 mm disc orifice. The maximum throw is about 15 m in still conditions.

Hand Operated Machines

In contrast to the extensive use made of hand compression sprayers, few organizations have adopted the lever-operated type, which is lighter but less robust. Unless a pressure regulating device is fitted, considerable variations in output rate can occur because of changes in lever operation, which is affected by operator fatigue. Operators have also found continuous pumping more tiring than the periodic pumping of compression sprayers and a further disadvantage is that they are unable to use both hands freely to gain access to difficult places or to steady long lances when spraying the undersides of high branches, one hand being required for pumping.

Pressure Retaining Compression Sprayers

Extensive use has been made of these machines because they maintain pressure and, consequently, less effort is required to operate them than nonpressure retaining sprayers. The high operating pressure also reduces the tendency to blockages. This type of sprayer can be rapidly charged by means of a detachable pump and maintains air pressure after the spray has been discharged; further pressure loss is prevented by a floating ball valve.

Two types of Colibri are still used in antitsetse operations, each using a different type of charge pump. The Leo-Colibri has its own detachable hand pump, whereas the Favori-Colibri machines are charged by means of a large manual or motorized pump which can charge several machines simultaneously. This can be time saving, especially where operators are working near tracks or in close proximity to each other, e.g. when treating narrow zones of riverine forest.

The detachable pump is another favourable feature of Colibri equipment because it reduces the weight and, consequently, operator fatigue, a matter of some importance in highly selective operations where accuracy is essential. In Uganda, where the Leo-Colibri is still in service, each sprayer operator is accompanied by a porter carrying both a reserve supply of insecticide and a charge pump. In Nigeria, a group of sprayers is serviced by a single mobile manual Favori-Colibri hand pump. The Colibri has been used also in the Central African Republic.

Non-pressure Retaining Sprayers

Non-pressure retaining sprayers have been widely used, often in preference to pressure retaining sprayers. This has been due to their lower cost, the nonavailability of suitable pressure retaining models and the need to stock fewer spare parts.

Some of the first insecticide trials in Uganda, were carried out with a Kent sprayer. In the past, this sprayer has seen brief service in East, West and Central Africa, but no reference has been found as to the model preferred. Three have been in production for many years with maximum working capacities of 4, 5, 9 and 13.6 1. The pump cylinder is of brass, tinned inside to prevent corrosion and is filled through a fairly small opening in the cylinder which can be closed with a plug. Another non-pressure retaining machine is the Warley, which is very similar to the Kent. This, too, is filled through an opening but sealed with a cap. In Uganda it was replaced by the Leo-Colibri before large scale spraying operations got under way, but its use in other countries is referred to by several authors. Like the Kent, there are three sizes of Warley sprayers of similar working capacities.

When the Leo-Colibri went out of production, a suitable pressure-retaining alternative could not be found. However, in both Uganda and Nigeria, the CP201 was then used, based on a Gloria machine with the same number. As with all non-pressure retaining sprayers, the pump remains in the machine while spraying is in progress. Like the Kent and Warley it was filled through an opening in the top of the cylinder provided for this purpose and closed with a plug. In its commercially available form, the CP 201 was found to be much less comfortable to carry than the Leo-Colibri. Its extra weight was a factor with the low position of the upper sling swivel support band as near the top of the tank as its curvature allowed. The commercial life of this sprayer was, however, short, and by 1971 it had been replaced by the CP 148.

The tank of the CP 148 is of stainless steel and the greater resistance to corrosion of this material may be an asset, although with proper care the corrosion of brass tanks is negligible. The tank provides greater tensile strength and the thinner sheet metal reduces its weight (7.4 kg). How well these thinner walls stand up to the falls which inevitably occur when treating vegetation on a rugged terrain remains to be seen. Unlike any of the machines referred to above, the CP 148 does not have a separate filling opening. Instead, the pump is removed and insecticide is poured in through that opening with the aid of the fairly large plastic funnel. This is firmly attached to the machine and is extended to surround and protect both the pressure gauge and the safety valve. Other models with smaller tank capacities which have been used in African countries are the Muratori, the Galeazzi and the Hudson x pert which also proved satisfactory.

A choice of either diaphragm pump or piston pump models, such as the Cosmos, is available and both have been utilized for spraying operations. There is one published record of the latter type, though these models have been used in Zambia. The Cosmos has a stainless steel 16 l capacity tank but the pump, which includes a paddle agitator, is of polyamid plastic. In contrast to other lever-operated machines, the Kestrel has a pump fitted to the outside of the tank.

A very different machine is the CP 3 which is now in service in Zambia and Uganda having replaced the Polclair. The polypropylene plastic body has a working capacity of 18 1 and mounted underneath it is a pump which is basically of the diaphragm type, but the upper diaphragm support has been so designed that it acts as a piston. The compression cylinder is not a separate component but is moulded in one piece with the pump body.

Pressure Regulators

When the fitting of a constant pressure device is to be recommended (16), an on/off tap to the outlet valve is necessary so that faults in the machine can be rectified without loss of spray liquid. Pressure regulators were once used extensively in Nigeria but were later dispensed with because spray deposit was more effectively controlled visually by the operators. A simple alternative to the pressure control valve is the disc flow regulator which has been incorporated in some sprayers.

Motorized Machines

The use of a small motorized pump is referred to in an account of spraying operations along the Malawa river on the Uganda/Kenya border in 1956. This type of pump was used from a dinghy and was capable of throwing a variable jet of spray to a distance of 8 m. It is believed that the 700 model spray guns were used in conjunction with the pump and, more recently, the banks of the same river were sprayed again from a boat. The insecticide was applied with similar guns but was supplied from a different type of motorized sprayer. This sprayer consisted of a high pressure diaphragm pump powered via a reduction gear by a 4-stroke petrol engine. Two delivery hoses were attached and an adjustable regulating valve gave a range of pressures from 10 to 30 kg/cm^2.

In Zambia the use of powered pumps has been the standard practice since large scale eradication campaigns were implemented. The equipment employed was a Platz spray unit mounted on Unimog model 411 tractor. The unit consists of a 3-stage piston pump driven by a pto shaft. Insecticide was suppled from an 800 1 tank mounted on the vehicle and was delivered through two hoses at a constant pressure of 30 kg/cm^2 to a pair of lances. These were held by two operators sitting on swivel seats at the back of the loading platform, but in rough conditions they walk alongside the vehicle. This type of equipment is, in comparison to manual methods, relatively expensive and wasteful (81) and is likely to be replaced by knapsack sprayers.

Gaseous Energy Equipment

Motorized Knapsack Mistblowers

These are used in most countries where tsetse control is practised. The model currently in use in Tanzania is the Motoblo Super 70 which closely resembles the Solo-Port 430. Both are equipped with 70 cc 2-stroke engines and 10 1 insecticide tanks.

Motoblo sprayers have been used in Tanzania since the early 1960's. Both the 60 and 90 models were also in use in Nigeria at about the same time. The model 60 was preferred to the latter because it was lighter and easier to handle. Further spraying operations with these machines have been described but they have since been replaced by the CP 40, which is now out of production. In spite of its smaller and somewhat unreliable engine, the CP 40 proved to be a reasonably effective machine. The plastic insecticide tank has a capacity of 10 1 and in Nigeria it proved effective for spraying persistent insecticides in swamp forest. A number of modifications were made to the CP 40, the most important being the replacement of the thin plastic air hose with a fabric-coated rubber hose, due to the plastic hose being very prone to perforation by thorns. This resulted in appreciable air loss and a consequent decline in the spray penetration.

Three KWH 77 machines were used in Uganda and they have been employed extensively elsewhere. Motorized knapsack mistblowers have also been employed in Rhodesia, but how long they have remained in service is not known.

Centrifugal Energy Equipment

The only record of equipment of this type used in control operations refers to the MEG. Briefly, this consists of a large centrifugal fan powered by a 420 cc air-cooled, 4-stroke engine. At the exit of the fan housing there is an aerofoil section containing a propellor which, powered by the air flow, rotates a wire cage at speeds of up to 20,000 rpm. Insecticide is fed into this cage through a nozzle and is emitted as small droplets which are carried away in the air blast.

Another type, the ULVA, is a light weight portable rotary atomizer. Insecticide is gravity fed to a nozzle in the centre of an atomizer consisting of two stacked corrugated plastic discs with serrated edges. These are rotated at speeds of up to 9000 rpm by a small 7-watt electric motor powered by a 12-volt battery. Extensive use is being made of the ULVA in crop protection work in the tropics, and in Nigeria it has been used to apply ULV formulations.

Smoke Generators

Some small scale control operations were carried out with smoke generators against G. palpalis (G. fuscipes) in Uganda (20, 47), but accurate dispersal of smoke under varying weather conditions was difficult. Efforts have been made to overcome these problems but this technique has not been used extensively.

Thermal Foggers

Field trials with the Todd Insecticidal Fog Applicator (TIFA), and its use in the treatment of small pockets of tsetse have been reported. The Swingfog has also been used, and larger versions of this machine, such as the SNIOO, are available for vehicle mounting.

Aerosol Dispensers

A problem in any tsetse control operation is the prevention of reinfestation, and an essential aspect of protective measures is the deflying of traffic by pickets. Tsetse flies which are carried by pedestrians, cyclists and private cars can be caught with nets, but this method is impractical for dealing with numerous flies which can be carried by a single lorry. In Uganda, small hand-operated sprayers are issued to pickets, and both gaseous and hydraulic energy models have been tried. The former, which are invariably cheap, insubstantial and intended for household use, are uneconomic because of their short working life and inefficient because of the wide variation of droplet sizes produced.

Good quality hydraulic energy models, such as the 2180, are more satisfactory. These are intended for indoor use but could be used in garage-type buildings erected at road pickets where lorries are treated.

Selection of Equipment

The diversity of types of spraying machines utilized in tsetse eradication work is largely attributable to the diversity of habitats occupied, not just by the genus as a whole, but by individual species. Ground spraying techniques have also been developed more or less independently in the countries where control is practised in a variety of ecological circumstances. Consequently, a variety of local factors have influenced the choice of equipment. The selection of machines is related to their spraying efficiency, cost, ease of handling, mechanical reliability, maintenance and availability of spares. Although tsetse control organizations carry out trials with different insecticides or formulations, there is rarely any mention of comparative trials of different spraying machines. Similar savanna woodland habitats of G. morsitans have been treated with spray guns and pressure-retaining compression sprayers in Uganda and by motorized knapsack mistblowers in Tanzania. In Uganda, persistent insecticides have been successfully applied in tsetse resting sites with a 700 spray gun. When treating savanna woodland and forest communities where the undergrowth is not too dense to pass on foot, or where access could be provided with a minimum of path cutting, a compression sprayer (preferably a pressure-retaining type) adapted for use with a spray gun has proved to be most practical. A pressure control valve is fitted to ensure even emission, and this is suitable for spraying trunks up to 3.5 m in height from a distance of 4 m. A much greater reach is often necessary when treating G. fuscipes infested vegetation from a boat. In such circumstances, a compression sprayer is inadequate, but a small portable, engine-driven, high-pressure diaphragm pump is used. With a motorized pump it is also possible to use an insecticide container with a much larger capacity than a compression sprayer, and this reduces refilling time to a minimum.

When dealing with thicketed or swamp forest habitats, access lanes are a costly item. Costs can be reduced by using suitable spray jets or motorized mistblowers which give better penetration in these situations.

Aerial Application Equipment

The use of insecticides has for some time been the accepted and main method of controlling tsetse flies and this is likely to remain so in the foreseeable future. Residual treatments of persistent insecticides (e.g. DDT and dieldrin) have been used both selectively and discriminatively by ground spraying methods. However, airspraying techniques have developed appreciably over the last few years, certainly to an extent where the use of aircraft to apply insecticides in antitsetse operations has progressed beyond the experimental stage. As far as the future is concerned, it seems likely that emphasis will be placed on the use of aircraft in many areas which are infested with flies, and it seems probable that any large scale control programmes will contain a significantly large serial input.

The principle reason for the choice of aircraft is the relative ease with which large scale operations can be carried out in comparison to spraying from the ground and also fewer trained staff required. Aircraft are able to apply spray to tsetse habitats which are inaccessible from the ground, and this either takes the form of placing the residual deposits in specific sites or by space spraying a wide area. Aerial applications of insecticides are likely to produce more general environmental contamination than ground spraying, except that much lower dosage rates are used, and therefore it is essential to incorporate ambient pesticide monitoring in any operational airspraying programme.

Application techniques by fixed-wing aircraft and helicopters have now reached the stage of development such that suitable equipment is available to meet most tsetse control requirements. Many large scale programmes to control tsetse flies are now based on aerial application of insecticides and, although they often involve a high financial commitment, aerial spraying may be less expensive than ground spraying in terms of unit area covered, depending on the insecticide selected and the amount of ground per unit area. This is partly accounted for by the need for fewer personnel and less logistic support in aerial operations than ground spraying and this often simplifies the operational planning. However, there are often difficulties in executing airspray operations because of the necessity to service and repair sophisticated equipment in remote areas and to adhere strictly to a schedule of sequential applications.

The Development of Airspraying Techniques

Research on application of insecticides from aircraft for tsetse control has progressed by two distinct and different methods. These methods are here referred to as "non-residual treatments" and "residual treatments". Although the use of aircraft in agriculture had its beginnings more than 50 years ago, for tsetse control, non-residual airspray treatments have been in use for almost 30 years and residual airspraying about eight years.

Non-residual treatments rely on sequential applications of very small droplets of a relatively concentrated insecticide, dispersed widely in savanna woodlands areas with the objective of applying spray droplets which will come into contact with tsetse flies, either in flight or in their resting sites . This method has been used extensively in Eastern, Central and Southern Africa. Residual spraying treatments from aircraft have been developed as an extension to the traditional ground techniques and adapted to aerial spraying of persistent insecticides to specific targets where tsetse are known to rest. These have been particularly suited to tsetse control in riverine vegetation, ecotones and drainage lines in West African conditions, especially where the tsetse habitat is restricted during the dry season.

Non-Residual Spraying Techniques

Aerial operations to control tsetse fly began shortly after the Second World War in South Africa, where Anson aircraft were fitted with extension tubes attached to their exhaust stubs. These devices were used to apply 5% solutions of DDT (102) to fly infested regions of Zululand. At about the same time in East Africa, work commenced on an experimental programme of aerial spraying to study methods of controlling tsetse fly. Work was carried out on an experimental basis and involved studies of insecticide formulations and spraying techniques with supporting physiochemical and biological assessments. Early experiments were carried out with war service aircraft fitted with crude devices, e.g. open pipes and thermal exhaust units which were used for the production of fine sprays and coarse aerosols.

Trials were ineffective in achieving control of fly but research continued throughout the years in improving application techniques, spray dispersal equipment and insecticide formulations. Eventually, improved fly control was achieved but only at high costs. The development of an improved thermal exhaust device fitted to a Cessna 182 aircraft which was capable of producing good aerosol characteristics was used to apply dieldrin and isobenzan which gave very satisfactory control of tsetse flies at reduced cost. In 1964, endosulfan was introduced for the first time and this experiment was shown to reduce costs appreciably.

These experiments pioneered the use of aircraft for the control of tsetse flies by non-residual spray methods and laid down the foundation of techniques which have developed over the last few years in Central and Southern Africa. In 1967, an airspray experiment was carried out to control tsetse fly in SW Barotseland, and this subsequently led to the commencement of a large scale spraying operation which was carried out the following year and covered an area of 1600 km^2. The techniques have been further refined in Botswana in 1973 where a twin-engine Piper Aztec aircraft fitted with a single Micronair unit has been used to control tsetse fly in the Okavango Delta.

Residual Spraying Techniques

Trials were carried out in Kenya using an experimental formulation of dieldrin applied by helicopter (Bell 470) and fixed-wing aircraft (Piper Pawnee 235) to control G. pallidipes. The insecticide was formulated as an invert emulsion and was applied from a Bi-Flon device. Spraying by helicopter proved to be prohibitive because of the high operational cost, but trials were continued using a fixed-wing aircraft.

The insecticide applied by this method was shown to penetrate the canopy and produce toxic deposits in the understorey thicket. A spraying technique was developed which consisted of applying a swath at intervals of 54 m across the thicket and this proved to be effective in controlling fly for at least eight months.

In West Africa, DDT, dieldrin and BHC as emulsion concentrate formulations have been applied primarily to clear riverine vegetation of G. tachinoides. Two applications at three weekly intervals were made from a Bell Helicopter (47G-4A 280 HP) fitted with boom and nozzle equipment. These areas have remained fly free for over four years. In 1971, discriminative spraying of dieldrin to vegetation usually sprayed from the ground was carried out in the Northern Guinea Savanna Zone (NGSZ) of Nigeria.

The area sprayed consisted of riverine vegetation, Isoberlinia woodland and ecotones around hills and in the sub-plains. A 20% dieldrin EC was applied at the rate of 1.5 to 2.0 kg/ha and, with isolated exceptions where fly pockets were missed, these areas have remained fly free for over three years. During 1972 spraying trials in the NGSZ, a helicopter fitted with boom and nozzle equipment was successful in controlling G. morsitans for several months using ULV formulations at 0.5 kg/ha and 1.5 kg/ha of dieldrin and endosulfan, respectively. Subsequently, trials using eight Ulva electrically operated rotary atomizers mounted on the boom of the helicopter indicated that tsetse fly was controlled and exterminated at 0.64 kg/ha and 0.8 kg/ha with ULV formulations of dieldrin and endosulfan respectively to about 10% of the infested area.

Equipment for Non-residual Insecticide Treatments

Open Pipe

The first trials carried out in East Africa employed an Avro-Anson XIX aircraft fitted with four spray tanks having a total capacity of about 800 l. Spray liquid was emitted under gravity from two wide pipes which extended about 35 cm below the fuselage of the aircraft. Emission rate could be adjusted by an Iris diaphragm which was situated at the end of each pipe. Adjustments were made by the pilot on the cockpit during the flight and spray atomization was achieved by the action of the slip stream which broke up the jet of liquid into spray droplets as it was emitted from the pipe. It was not surprising that the droplet spectrum from this rather crude device was fairly wide and the unit was subsequently discontinued.

Thermal Exhaust

One of the first devices to be used for the production of aerosol droplets was fitted to an Avro-Anson XIX aircraft. The coarse aerosol was produced by allowing the solution to flow into a modified exhaust system. The solution flowed through a narrow pipe 22 mm internal diameter and into the exhaust pipe at a distance of 50 cm from the centre line of the exhaust manifold. It then passed down along the exhaust tubes 10 cm in diameter, and was finally emitted vertically downwards into the slip stream about 30 cm below the trailing edge of the wing. The installation was based upon a design used for similar work against tsetse flies in South Africa.

In 1962, research was reopened to develop thermal exhaust units for aerosol production to enable studies to continue on aerial applications of insecticides for tsetse control. A device was developed for a Cessna 182, and this consisted of an Inconel tube 45.7 cm in length and 74.6 mm in diameter which was secured to the exhaust stub of the aircraft. The spray liquid was fed into the exhaust extension through a metering jet located outside the exhaust system. Another device consisted of fitting extensions to twin exhaust pipes, the extensions being two slightly curved stainless steel pipes 5.1 cm in diameter and approximately 60 cm in length.

A 340 l Sorensen spray tank was fitted below the fuselage of the aircraft and the system incorporated a 3-bladed, fan-driven Simplex centrifugal pump (2.54 cm). The flow of liquid was controlled by two diaphragm check valves, each fitted with a strainer and a D4 orifice disc. The liquid then flowed through two copper tubes 0.32 cm internal diameter to the outlet ports of the extension pipes.

The point of entry into the extension was 20 cm below the manifold where the temperature was estimated to be 550°C. Thermal exhaust units are currently in use in Zambia with the parastatal organization, "Rural Air Services", which has three Beechcraft "Baron" twin-engine aircraft with Sorensen external tanks of 360 l capacity.

Rotary Atomizers (Air Driven)

Micronair has produced a number of rotary atomizers over the last few years and these atomizers have been used on a variety of aircraft in numerous pest control operations throughout the world.

The physical characteristics of the spray emitted from the rotary atomizers are determined by the liquid throughput and the speed of rotation, but the droplet spectrum achieved is more uniform than hydraulic spray nozzles. The units are driven by a 6-bladed wind mill, whose blade angle and, therefore, speed of rotation can be adjusted between flights. Micronair equipment is used mainly for the ULV application of mists for the control of insects and diseases in crop protection and can be modified to produce aerosols in the range of 10–40 microns in diameter applied from small single-engine aircraft, provided that the insecticide solution is fairly volatile.

The latest unit to be manufactured in this series is the AU3000 Micronair rotary atomizer, and this has recently undergone trials in Botswana fitted to twin-engine aircraft to control tsetse fly. This has a smaller diameter cage (12.7 cm) than previous models, and when fitted to a Piper Aztec spraying at a speed of 260 km/h gave droplets of approximately 25–30ì vmd using insecticide solutions containing diesoline and Shellsol AB.

Equipment for Residual Insecticidal Treatments

Bi-Flon Apparatus

The Bi-Flon apparatus was designed specifically for the application of pesticides formulated as invert emulsions. Invert emulsions are produced when invert oils are mixed in water, the oil phase becomes continuous and the water phase dispersed.

The greater ratio of water to oil, the more viscous the emulsion. The use of inverts was designed to develop drift-free spraying, but the equipment and the method proved to be too complicated and the practice has now been discontinued. However, it is worthwhile mentioning some experiments which were carried out to control G. pallidipes and G. brevipalpis in the Lambwe Valley, Kenya because it may well have a bearing on future spray programmes using residual insecticide formulations.

Trials with a Bell 47G helicopter proved to be very effective in producing appreciable reductions in fly numbers, but the technique was too expensive and investigations were not pursued further. Trials were continued with fixed-wing aircraft and efforts to cut the cost by reducing both insecticide and spraying time were successful. Studies were made to evaluate the effect of spraying at intervals of 54 m, and it was shown that at least 60% of the area would have to be sprayed to achieve any satisfactory degree of fly control.

Boom and Nozzle Equipment

Residual spraying has been carried out in West Africa where helicopters have been in use to apply persistent deposits from the air in the Sudan savanna zone in Niger in 1969. An area mainly infested with G. tachinoides in dense riverine forest was sprayed twice with mixtures of DDT, dieldrin and BHC in an emulsion concentrate using a Bell helicopter 47G-4A fitted with boom and nozzle equipment.

In 1971, a trial covering 240 km^2 in the NGSZ in Nigeria was carried out with the same helicopter fitted with two spray tanks, each of 100 l capacity fixed to the right and left sides of the engine. A rotary pump which was belt driven from the engine was employed with 32 D3/25 nozzles at intervals of 50 cm. Applications were made at speeds of 32–40 km/h spraying to within 1–2 m of the canopy, and this gave an effective swath of about 20 m.

In larger broken areas of Isoberlinia woodland, swaths were sprayed at intervals of 200 m. Only about 10% of the area was treated according to the discriminative ground spraying technique developed earlier for ground spraying.

Dosage rates were reduced and techniques improved, and this was achieved by using fewer nozzles on the boom, thereby reducing the liquid throughput. In 1972, 12 of these nozzles were used to apply a ULV formulation of endosulfan in a vegetable oil. This ULV technique enabled helicopter applications to be made more efficiently in terms of insecticide used and in the overall cost of the operation.

Rotary Atomizers (Electrically Powered)

During the period 1972–73, spraying techniques in Northern Nigeria were further refined by employing modified Ulva atomizers and eight units were fitted to a Bell 47G helicopter. Three were fitted on each side of the boom and two behind the engine. Each atomizer was driven by a 24-volt motor and rotated at a speed of 9000 rpm. Liquid was carried from the boom through a restrictor to the atomizers and a constant flow was achieved.

The system allowed for multiple and single operation of the atomizers during the flight, and the spray techniques employed were similar to those used in earlier trials. The helicopter flew at 25–32 km/h over dense riverine vegetation and 40–47 km/h over open Isoberlinia woodland where spray swaths were made at intervals of 200 m. In 1973, an investigation was carried out to study the penetration of spray droplets when applied from electrically operated rotary atomizers fitted to a helicopter to a forest habitat in Ivory Coast. A Bell 47G-4A HP was used for the trials and carried two tanks, each 100 l, fixed to right and left sides of the engine.

The rotary pump, driven by a belt from the engine, supplied the spray system with liquid. The system consisted of eight electrically driven prototype Ulva atomizers which could be individually activated. These were attached at intervals of 0.5 m along a centre boom and two outer booms having a combined span of 10.5 m.

The 24V/60W atomizer motors gave a nominal rotation speed of 9000 rpm unloaded. Spray droplet sizes depended on throughput and this could be altered by changing the size and number of the liquid feed nozzles.

Experiments have also been carried out to study the effectiveness of several ULV insecticides applied by helicopters to control G. palpalis gambiensis and G. tachinoides in the area of the WHO Onchocerciasis Control Programme in the Volta River Basin.

The helicopter used was a Bell 47G-4A, 305 HP with a speed range of 32–80 km/h, and two rotary atomizers secured 6–8 m apart were used to apply ULV insecticide formulations. Two types were used: one type produced droplets between 20 and 30ì in diameter which were applied at a rate of 0.6 l/min; the other type, which had large apertures, produced droplets of 80–100ì diameter at 1.2 l/min.

The helicopter flew at a constant height of about 2 m above the canopy, but it was necessary for it to make side turns to follow the bends in the water course.

The powerful air current propelled downwards and backwards by the rotor carried the insecticide, which assisted the penetration of spray into the vegetation at ground level. In order to avoid losses due to convection near the surface of the ground, ULV applications were conducted for about two hours after sunrise.

Types of Aircraft Suitable for Antitsetse Operations

Small single-engine, fixed-wing aircraft are of limited use in tsetse control operations except possibly in the carrying out of experiments where relatively small areas need to be treated. In most instances, aircraft below 2500 kg maximum take-off weight would not have the range or weight carrying capability required for such programmes. It is probable that no single type of aircraft would be suitable for large scale tsetse control programmes and especially if a range of different tsetse habitats require treating.

Large flat areas can be satisfactorily sprayed with medium and possibly large multiengine aircraft, while small mountainous areas, often found in Eastern Africa, are probably more suitably treated by light twin-or medium single-engine aircraft.

Mountainous areas, areas difficult of access and especially winding riverine habitats are treated more effectively by a light helicopter. In general, the smaller types of fixed-wing aircraft used in tsetse control operations should have good pay load and range characteristics, be easy to control at low flying speeds and have good short take-off and landing (STOL) characteristics.

In addition, they should allow the pilot good forward and downward visibility and be corrosion proof against the effects of insecticide chemicals.

Tsetse control operations are often carried out in remote wooded areas where the opportunities for making a safe emergency landing (e.g. in the event of engine failure) are limited. For this reason, multiengine aircraft should be used whenever possible. Twin-engine aircraft should have a satisfactory single-engine performance at maximum take-off weight up to 2500 m density altitude.

With helicopters, important considerations are the forward and downward visibility for the pilot and the manoeuverability of the aircraft. The helicopter may be required to land in unprepared areas and a shielded tail rotor and high lift skids are valuable features. Any type of helicopter can be used in tsetse control operations, although machines with a maximum takeoff weight in excess of 1500 kg will probably be prohibitively expensive to operate.

Despite rising prices which have recently affected all aspects of pest control throughout the world, airspraying costs for tsetse control have actually decreased since techniques were first developed. This is attributed to a number of things but, primarily, spraying equipment design, better insecticides and improved application techniques have all played a major role in reducing costs.

Improvements in dispersal equipment have resulted in more efficient spray atomization which, as a consequence, has helped develop application techniques. Aircraft utilization has increased as a result of extending swath widths and high spraying speeds. Costs have also been reduced because of better pesticides which have been discovered and formulated as concentrates in aromatic oils containing higher percentages of active ingredients than was possible when field trials were initiated.

At periods of peak demand, the work output of aircraft can be extended by the employment of additional pilots over a seven day working week or by fitting high powered lights in the nose and on the wings of the aircraft which makes it possible to fly at night. Typical work output data have been derived for various types of aircraft which are, or may be, used in tsetse control programmes.

Navigation Aids

No ground markers are needed for helicopters in West Africa when treating tsetse in areas such as clearly visible drainage lines and vegetation interzones.

However, traditional methods of swath marking for aerial spraying in tsetse control operations over large woodland areas have been in use for some time, e.g. smoke fires, flags, balloons, marker bands, signal flares or lights.

The development of tsetse control techniques using fast aircraft, flying longer spray runs, wider swaths and night flying has made the development of more sophisticated navigational aids desirable. Several types of airborne guidance systems are under investigation:

One system is based on an airborne phase measuring device which measures the phase difference between the radio emissions from two or more ground stations.

The equipment is essentially a refinement of the Decca Navigation aircraft guidance system and has the disadvantage that ground transmitting stations are required sited at accurately mapped locations. Although easily portable ground stations are now available, these may be difficult to site correctly in remote areas.

Another system consists of a sensor unit which measures the inertia changes induced by altitude and acceleration changes in the aircraft during flight.

The inertial navigation systems (INS) are entirely self-contained within the aircraft and no ground stations are required. The main disadvantage of INS in vector control operations using light aircraft is the cost and weight of the equipment and the fact that the system is difficult to adjust to flying reciprocal tracks.

Medium and larger types of aircraft can be fitted with highly accurate electronic navigation equipment (e.g. Doppler, VLF), but ground markers are required to line up on approach and exit from large blocks, especially when spray runs are many kilometres in length.

In addition, these aircraft may be equipped with instruments to allow flying at night within close proximity to the woodland canopy. As weather conditions at night are often suitable for insecticide applications from aircraft, nocturnal tsetse control operations have been studied.

Cultural Control Practices

Cultural Controls: Making Your Site Unattractive to Pests

Cultural control methods include properly selecting and rotating crops, sanitizing and solarizing the soil, choosing the best planting and harvest times, using resistant varieties and certified plants, taking advantage of allelopathy, and intercrop-ping.

Crop Rotation

Certain pests are more common in some crops than in others. Rotating crops to different sites can isolate pests form their food source or can change the conditions pests must tolerate. If another site is not available, change the type of crops grown in the garden plot. Do not put members of the same plant family in the same location in consecutive seasons. For example, do not follow melons with cucumbers or squash.

This is also true for rotations using green manure crops, which add organic matter to the soil when they are tiled in before they produce flowers or seeds. Waiting two years to plant the same family of vegetable in the same location is the most effective rotation practice; however, yearly rotations can also be beneficial. Rotating annual flower plantings is also a good practice.

Sanitation

Many organisms responsible for disease and insect problems overwinter in plant debris such as shriveled fruit. Diseases on these shriveled fruit infect new leaves following spring. Removing crop residues, weeds, thatch, and volunteer.

Soil Solarization

A clear plastic sheet spread over the soil traps solar heat, which kills soilborne diseases, insects, nematodes, and many weed seeds. The treatment should occur during summer's high air temperatures and intense solar radiation. Keep the soil damp during the solarization process, and keep the plastic in place for several weeks. OSU Extension Fact Sheet EPP-7640 explains soil solarization in more detail.

Timed Plantings and Harvests

Many crops may be planted or harvested early to miss heavy pest infestations, while still achieving a full yield. Plant-ing earlier than normal may involve the use of cold frames or hot caps to protect seedlings form the weather while they get a head start growing. The crop then has a competitive edge over pests. Early planting depends upon the gardener knowing the emergence times and life cycles of the pests to be controlled.

Resistant Varieties

When buying seeds or plants, try to choose those with built-in resistance to diseases and nematodes. Sources for this information include OSU Extension Fact Sheets, seed catalogs, and plant and seed packages. It may be better to forego some production capability in favor of the increased pest resistance, if you must make such a choice. During the growing season, stressed plants can lose their resistance to pests, so be sure the crop has the water and nutrients it needs. When shopping for seeds and plants, check the labels for abbreviations similar to these, used to designate various types of pest resistance or tolerance:

A—Alternaria stem canker

ALS—angular leaf spot

ANTH—anthracnose

CMV—cucumber mosaic virus

DM—downey mildew

F—Fusarium (race 1)

FF—Fusarium (races 1 & 2)

MDM—maize dwarf mosaic

N—nematode

NCLB—northern corn leaf blight

PM—powdery mildew

SCLB—southern corn leaf blight

St—Stemphylium (gray leaf spot)

SW—Stewart's wilt

TLS—target leaf spot

TMV—tobacco mosaic virus

V—Verticillium.

Certified Plants

When they are available, consider buying plants labelled as "certified" or grown and inspected under sterile or quarantined conditions. Certified plants may cost more than others, but the certification guarantees they are free of diseases. Strawberries and potatoes are among crops which may be offered as certified plants.

Allelopathy

Allelopathy, a natural chemical interaction among plants, has been the subject of much research recently. Allelopathy refers to stimulatory as well as inhibitory properties. A living plant may release toxins, or in the case of decaying plant tissues, microorganisms may play a role in the release of the toxin. The microbes may also modify nontoxic compounds into toxic compounds. Black walnut trees and Johnson grass are among plants that have been shown to inhibit the growth of winter annual weeds and may offer some control of root knot nematode.

Intercropping or "Companion Planting"

The premise of companion planting is that certain plants repel insects, or attract beneficials that attack the insects. There is no significant data to prove the value of companion planting or intercropping, but it is thought that certain plants may produce substances which confuse insects, altering their impact as a pest. Some evidence also shows that planting flowers among vegetables attracts beneficial wasps seeking the flowers' nectar, and those wasps lay their eggs in the larva of certain pest species.

There is a popular but largely inaccurate belief among gardeners that marigold will control nematodes and other insects if planted among vegetables. Most marigold varieties do not have this capability. Only the French marigold (*Tagetes patula*) varieties, such as Nemagold.

Vegetables Families

Tomato Family: tomato, potato, pepper, eggplant.

Onion Family: onion, shallot, leek, chive, garlic.

Beet Family: beet, Swiss chard, spinach.

Cole Crop Family: cabbage, cauliflower, broccoli, brussels sprouts, bok choy, collards, kale, kohlrabi, mustard, radish, rutabaga, turnip.

Legume Family: bean, pea, cowpea, peanut.

Carrot Family: carrot, celery, celeriac, parsley.

Cucurbit Family: cucumber, watermelon, cantaloupe, pumpkin, squash, gourds.

Lettuce Family: lettuce, chicory, endive.

Green Manure Crop Family: hybrid sudangrass, buckwheat, soybean, cowpea, mung bean, garden pea, fava bean, ryegrass, rye grain, barley, oats, vetch, Austrian winter pea, clovers, greens.

Cultural Control of Drinking by Universal Moderation or Differential Access

The prospects for a new generation of cross-cultural hypotheses and analyses of drinking behaviour quietly improved a good deal over the past decade because of the emergence of a new and still largely untapped store of data. These data come from a growing series of broadly comparable social surveys focused on drinking behaviour and carried out in a variety of cultural settings. The trend began in 1970-1979 with, first, a four-country Scandinavian drinking survey carried out in Iceland, Norway, Sweden, and Finland (Makela, 1984, 1986) and, second, a three-culture study coordinated by the World Health Organization and carried out in community-level samples in Mexico, Scotland, and Zambia (Rootman, 1983). The WHO study — called the "WHO Community Response to Alcohol-Related Problems" project — spawned two additional comparable and nearly contemporaneous surveys in Canada (Smart, 1980) and in the U.S. (Roizen, 1981). More recently, surveys incorporating at least part of the WHO questionnaire have been done in Spain (Caetano and Martinez, 1987), in Japan (Clark, 1986), in a second region of Mexico (Caetano and Medina-Mora, 1986), and among Spanish-surname subgroups in the U.S. general population (Caetano, 1986). An August 1986 conference in Washington, D.C. — which drew participants from researchers involved in the original WHO project, from researchers at work on subsequent

WHO-like projects, and from others interested in this emerging resource—bespeaks the continuing vitality of this new tradition of research.

These studies offer a tonic to cross-cultural research not only because they are new and unfamiliar but because survey data bring a new kind of information to the territory, thus forcing new modes of thinking upon us. And although these survey data are not as richly textured as ethnographic reports nor as comprehensive in their coverage of the world's cultures as the Human Relations Area Files, they nevertheless offer the field a valuable new window on cross-cultural variation and a methodology that nicely complements the more familiar brands of cross-cultural data. For example, survey data almost invariably require an observer/analyst interested in cross-cultural variation to pay equal heed to variation *within* each cultural setting as well. Cross-cultural hypotheses trying to make their way in this sort of data environment must somehow manage to incorporate both cross-cultural and within-cultural variation into their explanatory agenda — no small duty.

Today, I would like to illustrate some of the charms of this sort of data by drawing on WHO-study data from the three original settings and the U.S. data, which in fact derive from a survey of Contra Costa County California in 1979. I have two goals. One is to illustrate how ordinary survey data can be used to examine structural and normative issues, issues that are usually reserved for other kinds of data and analysis styles. The other is to offer one or two conjectures on cross-cultural theory borrowing from these same data. Before we begin, however, a word or two should be said about the data sources. My survey data comes from seven separate samples: one in California, one representing the Lothian (Edinburgh) region of Scotland, two representing two kinds of communities — one urban and the other rural — in Mexico, and, finally, three samples representing an urban residential, periurban, and rural settlement in Zambia. Technically speaking, it is problematic to combine the samples drawn within a given country into a single sample, but that has been done nevertheless. Similarly, you should be aware that these data do not reflect the national populations of these settings but merely specific communities therein. Data, moreover, have been weighted everywhere besides the U.S. in order to bring the samples into equality across the genders.

Cultural Permission to Drink

Who respondents in Scotland, Mexico, Zambia, and the U.S. were asked to estimate much drinking would be "all right" for males and for females of four specific, hypothetical or test ages, 16, 21, 40, and 60. Four pre-coded answers were offered: (1) no drinking, (2) one or two drinks only, (3) enough to feel the effects but not drunk, and (4) "getting drunk is sometimes alright." The data I will be presenting are the frequency distributions for the whole samples in each setting; they tell us not the answers of males or of females regarding these age/sex statuses but instead the answers of the whole population about males and about females.

To begin, let us consider the normative patterns relating to the cultural control of access to drinking in general, a question we can address by initially collapsing together all three pro-drinking response categories. The Scottish and U.S. "permission" patterns were very similar. Both showed little difference in permission rates between the sexes (though females ran slightly below males), and both suggested an age-plateau form with high— and even nearly universal — permission granted for the three test ages twenty-one and over. Simple as they are, these curves seem to hint at three kinds of connections between drinking norms and the wider cultural environment — these links appearing as cultural premiums (1) on majority age, (2) on universalism or egalitarianism, and (3) on legalism. Concerning majority age, the normative pattern we see is consistent with a high cultural premium on a single, consensual, and officially-sanctioned "threshold" age of majority, one governing the right to drink as well as other adult prerogatives. The flatness of the curves across the three "adult" ages and the similarity of the curves for the sexes suggests a high cultural premium on universalism or egalitarianism. Finally, I see legalism hinted in the close agreement between popular opinion and official laws governing the right to drink. Legalism, in turn, suggests that the state holds and articulates the issue of cultural approval for drinking in the Scottish and U.S. settings.

The Mexican and Zambian permission patterns were quite different both from each other and from the universalistic and age-plateaued pattern of the Scottish/U.S. data. The Mexican data showed a substantial divide in cultural permission across the sexes. The Zambian data, on the other hand, showed wide variation in permission across test ages. Notice, for example, that all four female age-statuses in the Mexican sample are accorded lower rates of cultural permission than three of the four male age-statuses.

We have ordered the Zambian results from highest to lowest cultural permission level (first 40-year-olds, next 60-, next 21-, and, finally, 16-year-olds). Notice in this chart that the results follow a clear-cut pattern of age stratification: 40-year-old males and females ranked 1st and 2nd respectively in cultural permission levels; 60-year-old males and females ranked 3rd and 4th; 21-year-old males and females ranked 5th and 6th; and, finally, 16-year-old males and females ranked next-to-last and last. To be sure, no matter what the age-status, Zambian males were accorded higher levels of permission than females (even considerably higher for 21-year-olds), but more importantly, females in each age-status group were consistently accorded a permission level at least as high as that accorded males in the next-lower test-age group. Permission levels in the Zambian case, then, can fairly be said to be stratified primarily by age whereas permission in the Mexican case was stratified primarily by sex. In both the Mexican and the Zambian cases, moreover, permission to drink appears to involve a *more gradual or graded* cultural license in place of the *sharply divided* pattern at majority-age suggestion in the Scottish and U.S. permission norms.

The Mexican and Zambian normative patterns suggested neither the majority-age premium, nor the universalism, nor the legalism hinted in the Scottish/U.S. pattern. On the contrary, in both Mexico and Zambia the data reveal that the right to drink is much more strongly tied to one's ascribed status, or, in other words, to bases of social stratification more characteristic of traditional societies. Mexican and Zambian permission norms, then, seem to have 'stained-out' (to use Duster's [1983] evocative term) the ascribed-status bases of each society's broader social stratification systems, much as a biologist's dye stains tissue on a microscope slide. Mexico's gender-based and Zambia's age-based stratification systems are not in and of themselves particularly surprising. Mexico, of course, is well known for its machismo men and modest women, as Africa is known as home to many age-based systems of social stratification. Yet the striking difference between the two bases of stratification — gender and age — raise interesting questions. What are the chief differences between ascription based on age and on gender (La Fontaine, 1978)? How is it that permission-to-drink norms reflect the underlying architecture ascribed stratification in each cultural setting? What does the apparent parallelism between drinking permission and stratification tell us, on the one hand, about drinking and, on the other, about stratification? What are the chief differences in drinking practices, attitudes, and problems between universalistic and ascriptive social structures?

Finer-grain puzzles are also suggested in these findings. What does it mean, for example, that in Mexico *21-year-olds* (of either sex) were granted higher permission levels than like-sexed *60-year-olds*, whereas in Zambia the reverse was true — *60-year-olds* were granted higher levels 21-year-olds? Did the gradual, non-plateaued form of Mexican and Zambian permission norms imply that adulthood itself is likewise defined in an incremental or cumulative way? Why did the norms applying to 60-year-olds decline in levels of permission from those of 40-year-olds in Mexico and in Zambia as well as (if only in a hinted-at form) in the Scottish and U.S. data?

Permission to Drink and the Power of the State

If the permission-to-drink normative patterns hinted at different levels of involvement by the state in the regulation of drinking, then just such a relationship was unexpectedly lent support elsewhere in the WHO data, once more nicely illustrating the potential for survey data to address structural societal variables. As the study's title suggests, the WHO researchers were interested in community responses to alcohol-related problems. Its survey questions in this territory happened to include a series of four vignettes about hypothetical alcohol-related dilemmas. This series provided data on the perceived severity of alcohol-related problems in the community, on perceptions of such problems' perceived prevalences, and — most relevant here—on preferred kinds of social control responses. (Data on only three of the four vignettes will be considered here.) The first vignette described a man who had drunk so much that "he falls down on the road and cannot get up" (MANFALL); the second described a man who has on two occasions struck his wife when he was drunk (HITWIFE); the third described a man "who spends so much money on drinking that there is not enough food for his family" (NOFOOD).

Included in the battery of questions asked about each vignette were three questions asking respondents to evaluate the appropriateness of the involvement of various people or agencies in helping or controlling the man in question. For example, the questions asked (1) whether the man's relatives ought to help him, (2) whether a bystander on the street should do something, and (3) whether the police or appropriate authorities should be called in.

This series was introduced into the survey questionnaire in order to monitor norms and attitudes regarding the social handling of one or another sort of drinking-related problem. Yet its results also painted a clear picture of the relative ascendancy of the differential involvement of the state — represented by the "police/authorities" involvement — across the Scottish, Mexican, and Zambian samples.

Thus, it is worth briefly examining the different Patterns of control reported in these three setting. In the Zambian samples roughly 90 percent agreed that relatives should help in all three vignette situations. Concerning bystander involvement, 90 percent supported bystander involvement for the MANFALL situation and from 75-80 percent supported it for the HITWIFE and NOFOOD situations. Less than half the samples supported the involvement of the police or appropriate authorities in the MANFALL and HITWIFE situations, but almost three-quarters supported such involvement in the NOFOOD situation. Involvement by police/authorities ranked third — behind relatives and bystanders — for all three situations. The Scottish results were quite different. There, the vignette situations were seen as highly differentiated and sorted into two groups — the MANFALL situation, on the one hand, and the HITWIFE and NOFOOD situations, on the other. For all situations, involvement by relatives was supported by 80-85 percent of the sample. A like percent supported bystander intervention in the MANFALL situation, but involvement by police/authorities in this situation was supported by less than half the sample. The HITWIFE and NOFOOD situations — both of which involve third-party victims of the central actors drinking — were not seen as appropriate occasions for the involvement of bystanders by roughly three-quarters of the sample; in these, police/authorities was approved by over three-quarters of the sample. The patterns suggest an alternativity relationship between bystander and police/authorities involvements and a clear division of responsibility between the benevolent interventions (of a bystander) and the potentially punitive interventions (of police/authorities) — with state involvement contraindicating involvement by the citizen on the street. Clearly, the state's monopoly of police power and the differentiation and definition of its role is much greater in the Scottish than in the Zambian case. Interestingly, the Mexican pattern of responses appeared to fall between the Scottish and Zambian patterns.

Charts 14D, E, & F present the same data as 14A, B, & C this time contrasting each cultural setting with respect to a single vignette situation. Chart 14D reports the MANFALL situation, in which all three cultural settings reveal quite similar response patterns — high on relative and bystander involvement and low on police/authorities involvement. In the HITWIFE situation, however, Chart 14E throws the issue of state control in the Scottish setting into sharp relief. Here we see that the Zambian response pattern to HITWIFE is not much different from the response pattern to MANFALL. The Scottish pattern shows a clear preference for the involvement of police/authorities and only low support for bystander involvement. In the Mexican samples bystander and police/authorities involvements are roughly equally supported by about two-thirds of the sample. Finally, in the NOFOOD situation, the Zambian pattern shows more support for police/ authorities involvement, but such involvement appears to be added to high support levels for involvement by bystanders rather than substituting for it. In the Scottish and (though less so) in the Mexican case, as noted already, involvement by police/authorities is perceived more as an alternative to bystander involvement.

These patterns, then, give us a clear if only cursory glimpse of the normative expectations surrounding relatives, citizens on the street, and civil authorities or police in regard to exemplary problematic situations. They illustrate the marked difference in the expectations regarding state control across the three cultural settings examined and lend a measure of support to the idea that the growth of the state's power is probably structurally associated with the universalistic normative patterns evidenced in the earlier permission-norm data we examined.

Cultural Permission To Drink Enough to Feel Alcohol's Effects

Though the Scottish and U.S. norms governing access to drinking suggested a universalist form, it is well known, of course, that even in these settings drinking behaviour is strongly associated with (or marked by) gender and age variables. Indeed, the WHO data from this same study reaffirmed this long-standing truth. For example, fully 80-85 percent of Scottish and U.S. young males (18-29) reported drinking at least as often as once a week; whereas among older females (50+) the proportion drinking that frequently dropped to only 25-35 percent. Concerning such relationships, the draft WHO monograph noted that "when we examine frequent drinking, in all four cultural settings the distributions for males of all ages fall entirely above the distributions of their female compatriots." What light might the normative data shed the normative arrangements for heavier drinking? Let us return data from the age/sex drinking norms series, this time with that series' responses dichotomized so as to divide respondents who would permit "feeling the effects" or sometimes getting drunk" to a hypothetical status from those who suggested "no drinking" or only one or two drinks maximum. This dichotomy will tell us more about the normative structuring of higher-quantity drinking in each setting.

The Scottish and American patterns are interesting both for their similarities and their differences. First, it is apparent in both that permission to drink is not equivalent to permission to drink enough to feel alcohol's effects. In both sites, too, the permission distributions now taken on more shape — that is to say, deviated from the "flat plateau" age-structure evidenced in the earlier permission-to-drink curves. This suggests more of a culturally sanctioned age-based hierarchy with respect to higher-quantity drinking. In both settings and for both sexes, moreover, the apparent age-hierarchies evidenced in this new shape are the same: first, 40-year-olds, next 21-year-olds, next 60-year-olds, and, finally, 16-year-olds — this, a status with very low approval for higher-consumption drinking. Interestingly, the age-based hierarchies for both Scottish and U.S. females reveal a wider gap in permission levels between 40-year-olds and 21-year-olds than occurs between 21-year-olds and 60-year-olds. The same relative spacing is in evidence for Scottish males, if more weakly, but disappears for U.S. males.

Perhaps the most striking result of the Scottish and U.S. comparison, however, is the higher permission levels enjoyed by Scottish males in contrast with either Scottish females, U.S. males, or U.S. females. To take a More than 60 percent of the Scottish sample approved feeling alcohol's effects for a 40-year-old male, whereas the comparable approval levels for 40-year-old Scottish females was a little less than 50 percent, and for U.S. 40-year-old males and females roughly 40 percent. These results imply that the hierarchy of higher-consumption permission in the Scottish sample is ordered both by sex and by age whereas in the U.S. data — though small gender differences permission rates are evident — the permission hierarchy is geared to age alone.

The Mexican and Zambian data reveal different patterns. In the Mexican sites, permission to feel the effects is very low for females of all four ages. Indeed, females are by and large simply not permitted to drink enough to feel alcohol's effects — the highest approval level was won by 40-year-old females and that amounted to only about 15 percent. For approval ran from about 30 percent, for 21-and 60-year-olds to as much as 45 percent for 40-year-old males. Notice that although 21-year-old males received a ten-percentage point higher approval level than 60-year-olds for drinking, per se, approval levels for feeling the effects were roughly equal for the two ages. Sixteen-year-old males received approval from only about 5 percent of the sample, a figure that appears to be quite low until it is contrasted with the approval levels granted females. Zambia's pattern is different yet again. The striking finding here is the high levels of approval granted "feeling the effects" for 40- and 60-year-old males and females. On balance, the data for Mexico, Scotland, and the U.S. have shown that approval levels for feeling the effects among the three statuses 21-years or older run about half as high as approval levels for any drinking — a little higher for Scots, a little lower for the U.S. and Mexican samples. In the Zambian case, however, the ratio of approving effects to approving drinking jumps to roughly 70 percent. In the Zambian case, then, approval of drinking is almost equivalent to approval of feeling alcohol's effects — suggesting the possibility of quite different structural arrangements for drinking.

Some Speculations

It is intriguing to contrast the U.S. and the Zambian patterns with respect to permission-to-drink and higher-consumption permission. The U.S. norms suggested, of course, a pattern that permitted drinking universally to all adult statuses but imposed a normative obligation to restrict drinking to moderate levels. In the Zambian case permission to drink is a differentially allocated but the norms on consumption are much weaker. I have called these two models of normative control "universal moderation" (UM) and "differential access" (DA) — and it will be noted that they provided the title of my presentation today. My hunch is that these two models of normative control may carry broad implications for the structuring of drinking behaviour and offer an interesting line of connection between drinking behaviour, on the one hand, and other dimensions or aspects of culture.

The responses we have examined refer to "ideal" norms rather than actual conduct. We have seen, however, that even ideal norms in UM normative settings like the U.S. and Scotland pertain more to permission to drink per se than to heavier drinking. Drinking per se, then, appears to be symbolically associated with the universalistically defined rights and responsibilities of adulthood in these settings, but heavier drinking is symbolically associated with a separate stratum of cultural meaning, and one drawing on more traditional status arrangements.

I was particularly impress, for example, by the close agreement in heavier-consumption norms between U.S. and Mexican males. This result may, of course, be pure coincidence, but it nevertheless highlights the interesting possibility that drinking norms, particularly in less traditional cultures, occupy a point of tension or stress between two competing normative systems.

The difference between the U.S. and Scottish heavier-consumption norms is also interesting. If the structure of drinking's norms were tied firmly to the level of modernization in a society, then we would generally expect Scottish and U.S. norms at both permission levels to be the same. The standard indicators of modernization theory are, after all, quite broad and would give us no reason, I suspect, to call either the U.S. or Scotland the more modernized of the two. It follows, then, that the observed differences in U.S. and Scottish heavier-consumption norms suggests that modernization-type variables are not the only dimension of culture at work in shaping drinking's normative structure.

This, I think, is a telling possibility and may open up an interesting analytical approach to teasing out of survey data which aspects of culture, after all, are tied most securely to drinking. Just such findings, as it happens, can prove helpful in interpreting some of the other data collected in such surveys.

For example, indicators of "alcohol dependence" in the Scottish and U.S. samples revealed strikingly similar frequencies in the two samples across six of the seven items used. The seventh item, however, showed U.S. males and females reporting much higher frequencies than the Scots. This item, as it happens, asked if one had felt that he or she should stop drinking or cut down in the past year — a measure, perhaps, of both self-recrimination and the "dryness" of drinking norms. Could it be that the more permissive normative structure observed for Scottish men explains this curious difference in symptom frequencies. But if so, why do Scottish females differ from U.S. females on this measure as much or more than Scottish males differed from U.S. males?

Weed Management

Lesson Introduction

Explain to students that a great deal is already known about noxious weed control. One of the aspects of control is that a multi-pronged attack is usually taken with chemical, biological, and cultural control methods all being used either simultaneously or in sequence. Much research centres around development of plant specific controls, cost of controls, and reducing "collateral damage" that is inherent in the control methods utilized. Be sure to explain that the cost of fuel is always a consideration, whether it is fuel to run the spray rig and tractor, fuel to deliver the goats or bugs to the site and haul away the weeds you pull from the ground. This section of lessons has the opportunity to be expanded greatly. The two CD-Rom disks are very detailed and each section of each disk gives students a chance to study a different aspect of weed management. If students acquire current values for products and services, they can gain a greater understanding of the impact that weeds have on ranching, especially on cattle production.

Montana Education Content Standards & Benchmarks

Vocabulary

Acre – *noun*, 43,200 square feet or 1/640 of a square mile (5280' X 5280').

AUM – *noun*, animal unit month, amount of grass one cow eats in one month (about 600 pounds).

Biological control – *noun*, the type of living organism used to control the number of weeds, such as flies, sheep, and fungi.

Chemical control – *noun*, a process of controlling the number of weeds by spraying chemical herbicides on the plants. These chemicals are usually specific for broadleaf plants.

Cultural control – *noun*, a process of controlling the number and spread of weeds through plowing, pulling, mowing, burning, or irrigating.

Entomologist – *noun*, a scientist who studies insects and their life cycles.

Fungus – *noun*, a plant that lacks chlorophyll and obtains its nutrients from a host organism.

Gall fly – *noun*, a biological control that reduces the number of seeds produced by knapweed plants. Gall flies lay eggs in knapweed seed heads.

Herbicide – *noun*, a chemical used to control, suppress, or kill weeds by severely interrupting their growth process.

Infestation – *noun*, a troublesome or hostile spreading that threatens to overrun or inhabit an area in such large numbers that other organisms could be harmed.

IPM – *noun*, Integrated Pest Management, a program for weed control that uses all the available methods of control, rather than using just a single method.

Irrigate – *verb,* supply water to an ecosystem or crop land artificially to increase the productivity of the land.

Larva – *noun,* the immature, wingless, and often worm-like feeding form that hatches from the egg of many insects.

Section – *noun,* 640 acres or one mile X one mile.

Student Activities

1. Read and understand the "Montana Agricultural Statistics" pdf files. Research current costs and values of items mentioned and insert those present values and costs in place of those given in the handout.
2. Read the "Joe Smith Ranch Information Sheet" and calculate the results of the problems shown in questions 1-7 from. Show and label all work.
3. Read the "Chemical Weed Control Information Sheet" and answer questions 1-6. Show and label all work.

Cultural Control Strategies

Mulching

Mulches of wood chips, straw, grass clippings, paper, rocks, rubber, black plastic, or landscape fabric can be used to control weeds. Mulching is typically done on relatively small areas where the application can be monitored. Mulching works best on annual weeds and cannot control some perennial weeds because their extensive food reserves allow them to continue to grow up through the mulch. Wood chips are often used in forestry/wildland settings where woody weeds and undesirable vegetation can be shredded into wood chips and left onsite as mulch. In order to be effective as a weed barrier, the mulch should be 3 – 6 inches thick. Landscape fabrics are also used as mulches but are more commonly used on a smaller scale as they tend to be more expensive. These fabrics are heavy, sometimes woven and are made of polypropylene or other weather-resistant materials. Water passes through them slowly and many are made to prevent weeds from growing through them. When using organic mulches like wood chips, straw, or grass make sure the material comes from a weed free source and/or does not contain any weed seeds or fragments.

Burning

The most effective fires for controlling invasive plant species are typically those administered just before flower or seed set, or at the young seedling/sapling stage. Sometimes prescribed burns that were not originally designed to suppress an invasive species have that happy side effect. But in some cases, prescribed burns can unexpectedly promote an invasive, such as when their seeds are specially adapted to fire, or when they resprout vigorously. These prescriptions must be modified or other management actions taken to undo or reverse the promotion of the invader.

Most successful weed control efforts that result from burning are due to the restoration of historical (natural) fire regimes, which had been disrupted by land use changes, urban development, fire breaks, or fire suppression practices. Many prescribed burn programs are, in fact, designed to reduce the abundance of certain native woody species that spread into unburned pinelands, savannas, bogs, prairies, and other grasslands. Repeated burns are sometimes necessary to effectively control weedy plants, and herbicide treatments may be required to kill the flush of seedlings that germinate following a burn.

Spot-burning invasive weeds with a propane torch can be cheaper and easier than implementing a prescribed fire (permits are still required), but is only effective when the infestation is small. Spot-burning can be used to burn individual plants, groups of plants in a small area, or to ignite brush piles. Propane torches can be used in areas where there is little or no fine fuel to carry a prescribed burn, and can also be used to kill plants when conditions are wet.

Prescribed Burning and Herbicides

Some invasive species have underground storage organs that resprout vigorously after fire, and/or seeds whose germination is stimulated by fire. Some of these species may not be possible to control with fire, but some can be controlled with repeated burns and others may be especially vulnerable to herbicides after a burn. Resprouts or seedlings that are 1 to 3 months old are often especially sensitive to herbicides.

Timing of Burn

The timing of a burn can strongly affect the fire's impact on native and exotic plant populations. For example, in California's Carrizo Plain Natural Area, Meyer & Schiffman (1999) determined that warm-season prescribed burning (late-spring and fall) was most effective for reducing abundance of Mediterranean annual grasses. Native plant cover and diversity also increased significantly following warm-season prescribed burns. Winter burns, however, did not affect the abundance of native plants, and exotic plant cover was only moderately reduced.

Timing was also key in controlling smooth brome *(Bromus inermis)* and encouraging the growth of native grasses in Nebraska and Minnesota (Willson & Stubbendieck 2000). Timing prescribed burns so that they occurred at the time of tiller (aboveground lateral stem) elongation, yielded an immediate and persistent reduction in both tiller density and biomass of smooth brome.

Burning in Extensively Disturbed Areas

Not all burn treatments in wildlands are beneficial. When fires become too intense, crown-fires and death of native plants that typically survive fires can result. If temperatures are too hot, soil organisms and seeds, even those of species that require fire stratification for germination, may perish, and valuable soil nutrients may be volatilized or otherwise lost. In extensively disturbed areas of southwest Australia, fire actually enhanced the invasion of weeds along roadsides, and resulted in an overall decrease in the abundance of native species (Milberg & Lamont 1995). Schwartz & Heim (1996) reported that fire was at best moderately successful for garlic mustard (*Alliaria petiolata*) control in Illinois forests, and Luken & Shea (2000) determined that repeated prescribed burning had no significant effect on garlic mustard in Kentucky. In both cases, however, the burns were detrimental to native herbaceous species, reducing both density and richness. Even three years after the initial burns, native plant composition did not recover to pre-burn values.

Preventing Spread of Weeds

Keep all equipment, trucks, and engines clean of weed seeds. After each burn, and before moving to another site, be sure to clean (hose-off) all equipment, tools, and clothing used. This will minimize changes of carrying weed seeds directly to a new site where a fire might provide perfect conditions for their establishment.

Cultural Control of Established Dandelion in Turf

Dandelion can be a major weed problem in turf. It forms dense clumps which affects the footing and the overall playability of sports fields and golf courses. The texture and colour of its leaves does not blend harmoniously with the surrounding grasses. When in bloom, its yellow flowers are yet a further detractor from the aesthetics of the turf. Dandelion seed is spread on the wind, and as this seed can be carried great distances, prevention of new infestations is difficult. Once a dandelion becomes established and supported by its large taproot, they are even harder to control.

The foundation of cultural dandelion control in turf has always been to create a growing environment in which the turf has the competitive vigor to repel the establishment of new dandelion seedlings. However, when it comes to dealing with established dandelion, many of the cultural methods are abandoned, opting to solely rely on chemical control. As concerns over the use of herbicides for aesthetic purposes increases, the turf manager must explore new strategies and rethink the whole dandelion control process. In a previous study conducted by the Prairie Turfgrass Research Centre, mowing various species of grasses at a higher height was thought to reduce weed populations. A number of grasses resisted the development of weeds when mowed at 3" and there was virtually no weed development over the duration of the study. Aeration and topdressing are cultural practices that improve gas diffusion in soil which in turn helps to promote root growth. In addition, aeration physically removes thatch from the turf which also helps to improve the turf quality.

High quality turf as promoted by adequate fertility has also shown to reduce weed infestations. In addition, the two organic fertilizers, soybean and corn gluten meal, have shown to have weed suppression properties. This study was developed to determine the effects of various cultural strategies for the control of dandelion in turf.

Methodology

The effect of three cultural turf management practices on an established dandelion population was tested at the Prairie Turfgrass Research Centre on a site located at Olds College, Olds, Alberta. Two weeks prior to the establishment of the trial (July 8), a pretrial treatment of the herbicide Trillion was applied to the site in an attempt to suppress the dandelion infestation. Seven days after the herbicide had been applied the level of control was visually assessed. On June 22 nd the trial was established over the pretreated area and the location of each plot was identified and staked to allow for future reference. Test plots that measured 1x 1 metres were laid out on a turf site with a uniform stand of established dandelion. Each treatment was replicated four times within a three factor randomized complete block design.

To determine if soil aerification was a factor to consider when controlling established dandelions, half of the plots were aerified, while the other half was left untreated. A Ryan GA30 aerator, equipped with 5/8"hollow tines and set for a 3"x 3" spacing, was used to effectively open up the turf. The soil cores were harvested and removed from the site. USGA specification sand was spread over the aerified turf. Several passes with a drag mat were made to fill the core holes and to evenly distribute the sand over the aerated area. The second factor evaluated the effect of fertility on weed control. The individual fertilizers were weighed into plot sized lots and applied by hand using a simple shaker bottle. The plots received the initial application of fertilizer on the establishment date and a second application of the fertilizer was applied four weeks later. The mowing treatments were conducted on a weekly basis. This process was accomplished by first mowing the entire experimental area at the higher cutting height of 2½" (6.25cm), then using a push mower, selectively recut specific plots at the lower height of ¾" (1.9cm).

On August 31 st the trial was evaluated for the first time. The dandelions in each of the one square metre plots were physically counted. In addition the plots were visually evaluated for overall turf quality. Following National Turfgrass Evaluation Program (NTEP) protocols, three turf quality factors: colour, density and area cover were assessed. The colour factor subjectively evaluated the uniformity and intensity of the colour displayed by the turf. A 1 to 9 scale was used to rate each plot. Treatments which stimulated a uniform dark green colour received scores ranging from 6 for an acceptable colour to 9 for turf with outstanding colour. Plots that produced a lighter turf colour were scored lower. Density, the second quality factor, was subjectively evaluated for the impact of the treatments on the turf to produce more shoots and tillers. The 1 to 9 scale was again used to rate each plot. Treatments which stimulated a tightly knit turf received scores ranging from 6 for an acceptable density to 9 for superior turf. Treatments associated with a weak or thin turf stand were scored lower.

The final quality factor, area cover, subjectively evaluated the vigor of turf. Once again a 1 to 9 scale was used to rate each plot. Treatments which stimulated a thick competitive turf cover received scores ranging from 6 for an acceptable area cover to 9 for a superior area cover. Treatments producing a weak turf, affected by weed encroachment and/or the presence of bare patches, were scored lower.

Results

Herbicide Pretreatment

Seven days after the herbicide had been applied the level of control was visually assessed.

Classic signs of herbicide damage were evident throughout the trial site as the top growth of the dandelions appeared twisted and some upper leaves were beginning to turn brown. At this point, good control was expected. However, as dandelions counts ranged from 2229 dandelions per m 2 on the August 29 rating date, effective control was questioned.

Mowing Height

An analysis of the data indicates that there was no statistical difference in the dandelion population when the two mowing heights were compared. Even though the turf mowed at the higher height scored better than the turf mowed at the lower height for colour, density and area cover, it was not statistically different than the turf that was mowed at a lower height.

Turf Aeration

At the ninety percent confidence level the turf which was cored and top dressed had a significantly higher number of dandelions per square metre than the nonaerated turf. The turf colour and turf density of the nonaerated turf was significantly better from that of the aerated turf. There were no significant differences between the treatments for area cover.

Fertilizer

At the ninety percent confidence level the high rate of the controlled release synthetic fertilizer had fewer dandelions per square metre than the other fertilizer treatments. The turf colour stimulated by the high rates of the corn gluten and the controlled release synthetic was significantly better than the turf colour of the unfertilized treatment and the lower rates of either the soybean meal or the controlled release synthetic fertilizer.

Discussion

In the initial year of this study, the infestation of dandelion had remained very high. The pretrial herbicide treatment provided only short term suppression of the dandelion's top growth. With only a single set of data points it is still unclear as to which treatments or combination of treatments will have the greatest effect on the dandelion population.

Cultural Control of Herbicide-resistant Lolium Rigidum Gaud. Populations in Winter Cereal in Northeastern Spain

Lolium rigidum Gaud. is one of the most common weed species in winter cereals in Northeastern Spain. Herbicide resistance of *Lolium rigidum* Gaud. to chlortoluron and diclofop-methyl has been reported in Catalonia since 1995 (Taberner *et al.*, 1995). The problem has continued growing and more populations have been found resistant to tralkoxydim, fluazifop and chlorsulfuron (Taberner, 2005). The estimated affected area is still less than 0.5% of the 6 million hectares of winter cereal in Spain (Díaz and Gorrochategui, 2005) but the affected fields require costly management. Low cereal yields and selling prices have encouraged Spanish farmers to invest minimum efforts in their cereal crops spraying expensive graminicides only when the cheap ones failed. However, the current high cereal price situation may introduce new elements in this scenario modifying the prospects for herbicide resistance.

The great capacity of *L. rigidum* for developing resistance adapting to many different compounds has led to the development of multiple resistant populations in Australia (Powles, 1993) and to the existence of biotypes resistant to herbicides belonging to 10 different mode of action groups (Heap, 2005). Taking into account that climatic and edaphic conditions of some regions in Australia and Spain are similar, the herbicide resistance problem can potentially grow in Spain. In order to prevent and control this problem it is necessary to develop non-chemical control strategies, adapted to the local farmers' practices (Matthews, 1996; Gill and Holmes, 1997; Nalewaja, 1999). Out of these options, probably the easiest to be adopted in the Spanish context are delayed sowing, increasing weed density and ploughing. Stubble burning and crop topping are not authorised, pastures are not sown, and the resistance problem is probably not serious enough to justify modifying the external combine harvesters for seed catching as proposed by Matthews *et al.* (2004). If the problem grows, as it occurred in Australia, Spanish farmers will probably also adopt more non-chemical management strategies, improving these methods even years after the outcome of the problem, as exemplified with the seed burner presented by Matthews *et al.* (2004), thus showing that there is still interest in offering new alternatives to be combined with herbicides. Two main aspects of the seed biology of *L. rigidum* offer opportunities for its control. First, it is known that the viability of the seeds in the soil is short (Gramshaw and Stern, 1977; Taberner *et al.*, 1992) so that burying them by ploughing can reduce their viability rapidly. For seeds buried at 0, 5 or 20 cm depth, Taberner *et al.* (1992) found that one year afterwards only around 5% of the seeds remained viable in Spanish conditions. This behaviour opens the possibility of burying the seeds with a deep ploughing.

A second important characteristic of *L. rigidum* is that main germination occurs after rainfall in the beginning of the season (Gramshaw and Stern, 1977). Cheam *et al.* (1998) observed main emergence in the two weeks after crop sowing and up to 93% emergence concentrated in the 5-6 first weeks, while Matthews (1996) (using data of Heap, 1988) describes how 60-80% of the *L. rigidum* seeds emerged in the initial germination flush in Australian conditions.

This fact offers the opportunity of delaying the sowing date controlling the first *L. rigidum* emergence flush before sowing. However, depending on the climatic conditions, germination can also be staggered. In some cases, in Spanish conditions, there is a concentrated germination flush during the first 40 days after cereal sowing (Planes *et al.*, 1999) but in other cases the germination period can start 2-4 weeks after sowing and last 4-7 weeks (Recasens and Kuc, 1998). Sowing delay aiming a reduction in weed populations can thus be very variable, depending on the climatic conditions of that year. An important inconvenient of sowing delay is the possible yield decrease (Planes *et al.*, 1999; Anderson, 2007).

Lolium rigidum control caused by increased sowing density has been scarcely reported. Gill and Holmes (1997) stated that Australian farmers are usually not willing to increase wheat plant density above 100-150 plants m-2.

In Central Spain, Lacasta *et al.* (2004) tested the effect of sowing barley at three different densities (80, 160 and 240 kg ha-1) during eleven years in a monocrop situation finding that the intermediate rate (160 kg ha-1) generally yielded the most. In irrigated winter cereal, Moreno *et al.* (2002) recommended using more than 100 kg ha-1 in barley, as they did not find any significant yield increase in any of the three years after testing five different rates up to 200 kg ha-1 in the semiarid Spain. In different experiments at variable *L. rigidum* density, Izquierdo *et al.* (2003) did not find significant differences in barley yield for densities between 75 and 300 kg ha-1.

Liebman and Gallandt (1997) recommended combining several control techniques to have additive 148 *A. Cirujeda and A. Taberner/* Span J Agric Res (2009) 7(1), 146-154 effects, the so called 'many little hammers'. Similarly, Anderson (2007) proposed the dualistic approach of prevention and control, which is apparently effective in the semiarid steppe of the United States. However, little work on these techniques have been conducted in our area.

The objectives of this work were to test the effect of sowing delay, ploughing, increased cereal rate and the combination of these methods on *L. rigidum* control in winter cereal. The final aim was to provide new data on the effects of these methods, which should be combined in an Integrated Weed Management Programme to control herbicide resistant *L. rigidum*.

Material and Methods

Plant Material

Field trials were conducted on different herbicideresistant populations at different locations of the Lleida province. All experiments were conducted on commercial fields using the varieties chosen by the farmer and applying conventional fertilization. The resistance to herbicides was checked before treatment using a seed-based quick-test and confirmed in herbicide field trials. All the populations were resistant to chlortoluron, diclofop-methyl and tralkoxydim at the same time, excepting the population of barley in Sarroca, which was only resistant to chlortoluron. *Lolium rigidum* plant density in the control plots was high in all cases. Plots were placed perpendicular to the straw swaths, where this weed emerges predominantly (Blanco-Moreno *et al.*, 2004) so that *L. rigidum* distribution was similar between plots.

Cultural Control Methods

Sowing delay was conducted at five fields; mouldboard ploughing at four fields; the combination of both practices at two fields; increasing the cereal density and increasing the cereal density combined with sowing delay at one field. The experimental design was a split-plot with ploughing type as a main factor at Ferran 1, Verdú and Concabella. At Santa Maria and Ferran 2, the treatments were arranged as a randomised complete block design with two factors. At Sarroca, the experiment was a randomised complete block design with one factor. Elementary plots measured 4 x 5 m in all cases. Ploughing was conducted with a five-furrowmouldboard plough at Ferran 1, Concabella and Verdú at 15, 20 and 25 cm depths, respectively. At Ferran 2 a four-furrow plough was used inverting the soil approximately at 18 cm depth. Commercial ploughs were used in all cases. At Santa Maria, normal sowing density was 175 kg ha-1 corresponding to 439 seeds m-2 and high sowing density was 1.5 fold the normal rate, that is, 250 kg ha-1 corresponding to 627 seeds m-2. Normal densities in the area were used in the rest of locations.

Assessments

Lolium rigidum density was assessed at weed and crop heading (BBCH scale 56-59, BBCH Working Group, 1997) to describe the final effect on the weed population. Assessments included number of *L. rigidum* plants, number of *L. rigidum* ears and total fresh biomass in 0.1 m2 squares. Biomass was determined immediately after cutting in the field. Crop plant density per linear meter was assessed three times in each plot at heading at Santa Maria (28th May 2003), Verdú (17th June 2003) and at Sarroca (27th April 2004) to describe the effect of the cultural methods on the crop.

Data Analysis

Data were transformed when necessary to satisfy normality and variance homogeneity following the indications of the Box-Cox Transformation (Bowley, 1999) using $(x + 1)0.5$ for most cases, excepting weed weight per plant at Ferran 1 where $x0.5$ was used. Analyses of variance were performed using the SAS GLM procedure (SAS Institute, 1991) for each location separately considering the different models. Least significant differences (LSD) were calculated for pairwise comparisons. Mean values are shown as least square means (LSM) and differences between means were analysed on the basis of ordinary t-tests considering $P < 0.05$ in the case of significant interactions.

Results

Mouldboard Ploughing

Mouldboard ploughing reduced significantly *L. rigidum* density at Concabella and Verdú. This difference was particularly remarkable at Concabella, where ploughing reduced 96% weed density. Although ploughing reduced 55% weed density at Ferran 1, this difference was not significant. Similar results were obtained for weed biomass in these three sites. No significant differences in densities and biomass were observed at Ferran 2. The lack of efficacy of ploughing at this site is probably due to the high stoniness at this location, which impeded deep enough ploughing without damaging the tool. Ploughing did not affect wheat density at Verdú, the only site where this parameter was measured.

Sowing Density

Increasing sowing density had no effect on crop density and on *L. rigidum* density and biomass at Santa Maria, the only site where this comparison was conducted.

The Combination of Different Techniques

The combination of sowing delay with ploughing or with sowing density did not improve the effect of the individual treatments significantly at any of the three sites but had some effect at Ferran 1, where *L. rigidum* density was reduced. At Ferran 1 and Verdú, ploughing was the most important control method. At Santa Maria, the main weed reduction was due to sowing delay.

Concerning the effect of these techniques on the crop, at Santa Maria, increasing the sowing rate at the second sowing date gave significantly higher wheat density than the other treatments. However, no effect was observed at Verdú on the crop due to the combination of both techniques.

Discussion

Efficacy on weed plants due to ploughing was not considerable at Ferran 2, around 50% at Ferran 1, 81% at Concabella and 93% at Verdú. These values are similar to those obtained by Recasens *et al.* (2001b) (90 and 81%) and by (Recasens *et al.,* 2003) (65 and 92%) in Northeastern Spain, a bit higher to those obtained by Gill and Holmes (1997) (13 and 64%) and a bit lower to those found by Matthews *et al.* (1996) (98%), in Australia. The variation between the experiments suggests that ploughing needs a precise implementation. It should be mentioned that a correct ploughing can be difficult in stony soils like the one in Ferran 2. High fuel consumption justifies its use only in extremely highinfested fields or if cereal price is high enough to compensate the ploughing costs.

Trial results confirm that climatic conditions in the time gap between the normal and the delayed sowing date are crucial for efficacy and should allow maximal *L. rigidum* emergence. Efficacy on weed plants was not considerable at Concabella and Verdú, 35% at Santa Maria, 54% at Ferran 1, and 88% at Sarroca. The highest efficacy was achieved in the field with the highest

Sowing Delay

Sowing delay affected significantly *L. rigidum* density at Ferran 1, Santa Maria and Sarroca. At Ferran 1 and Sarroca, rainfall between both sowing dates allowed new weed emergences, which were mechanically controlled, prior to the second sowing date. At Santa Maria, although little rainfall fell between both sowing dates the soil remained moist due to the abundant rainfall in the months before sowing and to the foggy days, allowing weed emergence. Total weed biomass was also significantly lower in the second sowing date at these three locations. In the other two trials located at Verdú and Concabella, there were no differences in either plant weed density and weed biomass between the two sowing dates. At these two locations too little rainfall fell before the first sowing date and between the two sowing dates to allow *L. rigidum* emergence. The overall efficacy of sowing delay in these trials was lower than the control obtained by ploughing. The effect of sowing delay on crop density was not statistically significant at Verdú, but at Santa Maria and at Sarroca significantly more crop plants established at the second sowing date ($P < 0.05$). At Santa Maria, no significant differences and even no clear tendencies on *L. rigidum* were detected due to increased sowing rate, so that probably wheat plants were not more competitive when sown in 1.5-fold the normal density.

A possible explanation is that the field at Santa Maria was irrigated, consequently, water was not a limiting factor. Increasing sowing density could have a more important effect on weeds in rainfed environments as found by Izquierdo *et al.* (2003) in a barley crop infested with *L. rigidum* who describe a lower weed biomass for the same number of weed plants. Taking into account the yield results of Moreno *et al.* (2002) and Lacasta *et al.* (2004), probably it is not worth to increase the sowing rate more than 160 kg ha-1 in these semi-arid environments.

The Combination of Different Techniques

The theory of using 'many little hammers' to suppress weeds (Liebman and Gallandt, 1997) was partly confirmed at Ferran 1 and also by Recasens *et al.* (2001a). However, the experience of Anderson (2007) in the semiarid steppe of the United States shows that more than two cultural control tactics may be necessary to be effective enough. In the semi-arid conditions of Spain, lack of water in the drylands hinders to integrate other tools proposed by Anderson (2007) as crop rotations of cold-and warm-season crops. In line with the experiences in his situation, no-till could be also used as an effective tool under Spanish conditions. Because seeds remain on the soil surface, they are vulnerable to specific mortality factors that act on surface lying seeds. Seed mortality could contribute declinining the weed seedbank. For example, Baraibar *et al.* (2008) found that predation by harvester ants (*Messor barbarus*) of *L. multiflorum* seeds was significantly higher in no-till than in conventional tilled fields. In case of combining sowing delay with herbicide use, an additional benefit of delayed sowing is that later emerged *L. rigidum* plants are weaker and probably more susceptible to herbicides or to other control strategies (Recasens *et al.*, 2001a).

The starting situation of the fields in the trials described in this work is of highly-infested fields with resistant weeds, which could not be controlled with the conventional herbicide use. In this context, weed density after combining two cultural control methods was still very high ranging between 35 and 450 *L. rigidum* plants m-2 depending on the field. Thus, in highlyinfested fields with herbicide resistant *L. rigidum,* it will probably be necessary to combine these cultural methods with other more drastic tools as leaving the field fallow for one year, depleting the seed bank by repeating cultivation as suggested by Gramshaw and Stern (1977), or combining them with the use of those herbicides that are still active on the populations. As commented previously, the new high cereal prices will probably favour the use of more expensive active ingredients, which should be used correctly in combination with other techniques without relying exclusively on them.

Taking into account that delayed sowing date can have negative consequences on yield due to lack of rainfall and that the effect on *L. rigidum* is more variable, mouldboard ploughing can be a more secure strategy for its control than sowing delay, with more constant effects on weed plant reduction and with a lower impact on yield. The main problem of this technique is the high fuel consumption, which needs to be compensated by the cereal price and by the *L. rigidum* control. Taking into account the fast decline of the seed bank of this species, it is probably useful to plough every 2-3 years, only, to reduce the resistant populations. As the highest reductions were achieved by combining ploughing with sowing delay, it is probably worth to practice both, though high densities of *L. rigidum* will require additional control methods.

Development of Cultural Control Methods for Pests of Leafy Vegetables

This project aimed to evaluate whether or not certain plantings of non-crop plants can assist with pest control in leafy-vegetables such as lettuce and celery. It also assessed the potential of other management practices such as addition of composted fowl manure and using flowering weeds to augment populations of predatory and parasitoids insects and mites. The target pests were primarily aphids and thrips. These are important groups for leafy vegetables as they include lettuce aphid (*N asonovia ribis-nigri*) and western flower thrips (WFT) (*Frankliniella occidentalis*), both of which are resistant to a range of insecticides.

WFT is also an important vector of tomato spotted wilt virus. Strip plantings of cereals will attract certain species of aphids which may be pests in some crops (such as wheat or barley) but are not pests in leafy vegetables. However, predators of cereal aphids such as brown lacewings (*Micromus tasmaniae*) are predators of pest aphids and other insect pests in leafy vegetables. We aimed to find out whether or not strip plantings of cereals within horticultural crops can (i) increase the populations of beneficial species and (ii) increase the level of control of pests. Mites, staphylinid beetles and predatory thrips are known to prey on thrips and this project investigated how populations of these groups could be encouraged. It has been noted recently (Beaulieu and Weeks 2007) that there are many soil-dwelling predatory mites that have potential as biological control agents in agriculture, but there has been extremely little research on this group. The use of composted fowl manure was seen as one potential means of achieving increases in resident, soil-dwelling predators. The sequence of events is hypothesised as follows:

1. The application of composted fowl manure will result in more sciarid flies.
2. Predatory mites and beetles that prey on these flies (as eggs and larvae) will be advantaged
3. The predatory mites that are encouraged feed on WFT
4. Therefore, improved control of WFT can be achieved by enhancing generalist predators including predatory mites.

There have been many studies on the provision of pollen and nectar sources to augment populations of biological control agents, particularly hoverflies. There are also some studies where "insectary plantings" have been used to provide an in-field resource to maintain populations of beneficial insects (Colley and Luna 2000). Examples of this type of approach from Australia include the use of strip plantings of lucerne in cotton (Cotton CRC 2005), grasses in citrus (Smith *et al.* 1997) and buckwheat in vineyards (Bernard *et al.* 2006, Berndt *et al.* 2002). In this project we conducted preliminary observations on the potential for existing weeds, particularly flowering aniseed weeds, to provide the same resource for beneficials.

4

Soil Bacteria and Actinomycetes

What they are and a Few Interesting Facts

Bacteria are minuscule, one-celled organisms that can only be seen with a powerful light (1000X) or electron microscope (we're talking TINY). They can be so numerous that a pinch of soil can contain millions of organisms.

Bacteria are tough—they occur everywhere on earth and have even been found over a mile down into the core of the earth. Bacteria can be classified into five functional groups. Autotrophic (literally, self-feeding) bacteria are photosynthetic. They are the primary producers.

Decomposers consume soil organic matter, plant litter, and simple carbon compounds, releasing the nutrients in these substances for use by living plants.

Mutualists, such as nitrogen-fixing bacteria, form associations with plants and help them absorb nutrients. Pathogens are the bad guys— they cause disease in plants.

The last group, the chemolithotrophs (literally, chemical and rock-eating) obtain energy from minerals rather than from carbon compounds.

Bacteria are common throughout the soil, but tend to be most abundant in or adjacent to plant roots, an important food source.

Actinomycetes are a broad group of bacteria that form thread-like filaments in the soil. They are responsible for the distinctive scent of freshly exposed, moist soil.

Why they are Important

Bacteria are important in the carbon cycle. They contribute carbon to the system by fixation (photosynthesis) and decomposition. Bacteria are important decomposers in grassland environments. Actinomycetes are particularly effective at breaking down tough substances like cellulose (which makes up the cell walls of plants) and chitin (which makes up the cell walls of fungi) even under harsh conditions, such as high soil pH. Some management activities, particularly those that change nutrient levels in the soil, can shift the dominance of decomposers from bacterial to fungal. When one group becomes dominant where it shouldn't be, there is also a shift in the rest of the system. The shift from bacterial to fungal dominance, for instance, can enhance the conditions favoring weed invasions on rangelands.

Bacteria are particularly important in nitrogen cycling. Free-living bacteria fix atmospheric nitrogen, adding it to the soil nitrogen pool. Other nitrogen-fixing bacteria form associations with the roots of leguminous plants such as lupine, clover, alfalfa, and milkvetches. Actinomycetes form associations with some non-leguminous plants (important species are bitterbrush, mountain mahogany, cliffrose, and ceanothus) and fix nitrogen, which is then available to both the host and other plants in the near vicinity. Some soil nitrogen is unusable by plants until bacteria convert it to forms that can be easily assimilated. Some bacteria exude a sticky substance that helps bind soil particles into small aggregates. So despite their small size, they help improve water infiltration, water-holding capacity, soil stability, and aeration. Wait! Aren't there also "bad" bacteria? Yes, there are, but some soil bacteria suppress root-disease in plants by competing with pathenogenic organisms. The key is in maintaining a healthy system so that the good guys can do their work.

Bacteria are becoming increasingly important in bioremediation, meaning that we (people) can use bacteria to help us clean up our messes. Bacteria are capable of filtering and degrading a large variety of human-made pollutants in the soil and groundwater so that they are no longer toxic. The list of materials they can detoxify includes herbicides, heavy metals, and petroleum products.

Some Soil Bacteria Protect Soybeans from Root Diseases

Brian McSpadden Gardener, a plant pathologist with the Ohio Agricultural Research and Development Centre, is studying biological control of plant pathogens using beneficial bacteria that colonize plant roots. The goal of the research is to identify the distribution of soil bacteria across Ohio to determine which soybean fields would be ideal candidates for biological control applications.

"We know that some bacteria can promote plant growth but what we don't know is how much of those beneficial bacteria are in individual fields or how widely distributed they are across any defined geographical area," said McSpadden Gardener. "The ultimate goal is to find out if there is a way to assay a field and say okay, there are very low populations of beneficial bacteria here, so this would be an ideal spot to apply biological control."

In his work, in collaboration with Ohio State University Extension soybean agronomist Jim Beuerlein, McSpadden Gardener has identified two bacterial genotypes specific to Ohio soils that produce a natural fungicide known as DAPG (2,4-diacetylphloroglucinol). DAPG-producing bacteria that colonize soybean roots protect the plant from a wide range of root rot diseases, including Phytophthora, Pythium, Rhizoctonia and Fusarium.

"What we are doing is isolating different representatives of these DAPG-producing bacteria and treating seeds with them to see how well they colonize soybean roots and how well they protect roots against diseases," said McSpadden Gardener. "Our goal is to improve stand and increase plant yields by applying these DAPG-producing bacteria to seed as natural fungicides."

The DAPG compound works much like a chemical seed treatment fungicide in that it diffuses away from the plant roots, protecting the plant from fungal pathogens. But, unlike chemicals that lose their effectiveness once the plant grows out of its juvenile stage, DAPG-producing bacteria provide the plant protection from germination up to harvest.

"With synthetic chemicals, once that plant grows out of its sphere of influence, the fungi can sniff out the roots and attack," said McSpadden Gardener. "The idea behind biological control is that if you put the protective bacteria on the seed, as the plant germinates, the roots get colonized by the beneficial bacteria and they keep reproducing as the roots grow so more of the roots are protected more of the time. There's no other way to protect plants like that with a seed treatment."

During the two-year study, funded by an OARDC Seed Grant and the U.S. Department of Agriculture National Research Initiative, the researchers found that at 10 soybean sites throughout Ohio only about half of the plants harbored DAPG-producing bacteria. McSpadden Gardener said soil type, bacterium genotype, soybean variety, weather and production management are just some of the factors that determine the level of beneficial bacteria that are present to colonize plant roots.

"That's why it's so important to identify genotypes that are indigenous to Ohio," said McSpadden Gardener, who is also studying bacillus species that may prove effective in Ohio against soybean diseases. "A non-indigenous bacterium is not familiar with the Ohio soils or the environment, so it may not be as effective."

Seedling diseases cause a loss of about 3/4 of a million tons of seed annually, accounting for over $150 million in U.S. losses. Beneficial bacteria may be used as natural, effective and economical disease controls because of the unique relationship plants have with soil microbes. Between 5 and 10 percent of the carbon a plant fixes is released into the soil providing food for bacteria that colonize plant roots. Managing those plant-microbe associations can improve plant health.

Soil Bacteria's Role in Healthy Soil

Bacteria are the unsung heroes of gardening. Without bacteria, gardeners would not have the rich, loamy soil in which plants grow without bounds. Bacteria are single-celled organisms that chemically digest organic matter in soils into smaller nutrient components in forms available to plants. There are hundreds of thousands of different kinds of bacteria, and many types of bacteria can digest hundreds of different forms of organic matter into humus. The bacteria are able to do this because they can produce many different types of enzymes to digest different compounds.

Bacteria in Soil vs. Bacteria in Compost

Each type of bacteria works best under certain conditions. In a cool compost pile, or in garden soil, bacteria that thrive in cool conditions will form the bulk of the bacterial population. In hot compost, other bacteria that can survive the high temperatures dominate. When considering a bacterial soil conditioner, keep in mind the way in which you plan to use the conditioner. Understand the bacteria you are adding, if you are adding bacteria as soil conditioners, so that you add the right kind of bacteria. Once conditions change drastically, bacteria that cannot function in those conditions will perish. (Bacteria are not very mobile, unless they have water to float along in or wind to carry them.) The upside of bacterial immobility is that they will stay and multiply in place as long as conditions are favorable.

Bacteria for Soil Structure Improvement

Using a bacterial soil conditioning program not only helps with nutrient availability, it also helps to improve soil structure. Soils with poor structure benefit as bacteria breaks down soil compounds and the soil re-aggregates. Spaces for air and water will open up, and the structure of the soil will become more uniform.

Add Beneficial Bacteria for Oxygenation

Well structured soils provide plants with necessary oxygen in the root zone. Plants use carbon dioxide for photosynthesis, but they use oxygen for respiration, which is the process whereby plants break down stored sugars and starches to use as energy for growth. They get their oxygen by absorbing it in the root zone. Soils with good structure have plenty of spaces for oxygen. Soils without structure or organic matter generally do not have enough oxygen. When the plant cannot successfully undergo respiration, it cannot grow well.

Not all Bacteria is Good Bacteria

There are beneficial bacteria and there are bacteria that spread diseases. For bacteria to aid in soil conditioning, they need to be beneficial. Not just any bacteria will do. You can't really open your refrigerator, grab something old and throw it in the garden! You need the type of bacteria that will function well in your soil, in your climate, to decompose organic matter. Good bacteria does something other than break down organic matter. Good bacteria in soil is instrumental in keeping harmful fungi and viruses at bay. Be careful about your source for garden bacteria. Many will sing the praises of compost tea for adding bacteria to your soil, but these teas can also add pathogens that will harm your plants. If adding bacteria to your soil, it is best to go with a reputable source that can provide beneficial bacteria that will produce results.

Soil Components and Microorganisms

The oil is composed of five major components, such as inorganic matter, organic matter, soil air, soil water and soil organisms. The inorganic matter is derived from the weathering of parents rocks; air and water are derived from the atmosphere and are present in the soil; the organic matter is derived from the decomposition of organic matter; and the soil organisms constitute the most vital component of soil body. The soils, in fact are excellent culture media for the growth of various organisms. It is convenient to divide soil organisms into two major categories. First category refers to all those organisms which Can be studies satisfactorily only with a microscope; these are the microorganisms which include bacteria, cyanobacteria, microfungi, microalgae and protozoans. The second category refers to the remaining organism represented mostly by animals which can be studied by naked eye, and some are even quite large.

The greatest amount of microorganisms (1,000,000 per cu cm) is found generally in the top layer of the soil up to a depth of 5 to 15 cm. In deeper layers (1.5 to 5m) individual microbes are found. However, they have been discovered at a depth of 17.5 m in coal, oil, and artesian water. Due to nutrients and mixture in the soil, the number of microorganisms in 2 gm of soil reaches a colossal number from 200 million (bacteria) in clay to 5,000 million in black soil.

Bacteria That Infect Plants

Most people think of bacterial diseases affecting animals and humans but plants can also succumb to bacterial infections. Bacteria can enter through cuts our other areas of damage that occur in the leaves or stems due to wind bending the plant, objects hitting the plant, aphids sucking the plant sap or animals grazing on it. The symptoms then depend on which internal tissues and structures the infection takes a hold in.

Although thousands of species of bacteria can cause disease in animals and humans, a much smaller number, probably about one hundred, are able to infect plants and damage them. Bacterial plant diseases are generally due to bacilli, bacteria that are shaped by rods, some of which are gram negative bacteria and some of which are gram positive. The majority of plant diseases due to bacterial infection occur in parts of the world that have a tropical weather pattern.

How Bacteria Damage Plants

Once a species of pathogenic bacteria gains entry to a plant it causes one of four main problems that leads to a less than healthy plant. Some bacteria produce enzymes that break down the cell walls of plants anywhere in the plant. They do this to break open to cell to gain access to the nutrients inside but the plant cells affected die quickly, causing parts of the plant to start rotting. This is why plant diseases that develop due to cell wall degrading enzymes are generally known as a type of 'rot'.

Some bacteria produce toxins that are damaging to plant tissues generally, usually causing early death of the plant. Others produce large amounts of polysaccharide sugars that have long chains and are very sticky. As these travel in the water carrying vessels, the xylem, they block the narrow channels, preventing water getting from the plant roots up to the shoots and leaves, again causing rapid death of the plant. Finally, some bacteria produce proteins that mimic plant hormones. These lead to overgrowth of plant tissue and tumours form. These grow rapidly, taking up valuable nutrients and energy resources that the plant would otherwise use for its own tissue growth and it becomes weakened and susceptible to attack by other pathogens such as fungi.

Types of Bacteria That Infect Plants

One of the main types of bacteria that cause plant disease are the Proteobacteria. This large group of bacterial species include bacteria that cause disease in humans, such as Vibrio cholerae, which causes cholera, and Helicobacter pylori, which is associated with stomach ulcers. Some species of Proteobacteria live in soil where they fix nitrogen from the air, forming useful compounds that plants use in their growth. One species, Xanthomonas campestris, which affects plants from the Brassica and Arabidopsis groups causing black rot. The disease symptoms progress from discoloured leaves to extensive wilting of all of the stems and leaves, with all the leaves eventually turning yellow. Parts of the plant start to die and rot and the plant dies completely shortly afterwards.

Xanthomonas campestris causes its effects on the plant by producing large amounts of polysaccharides that block the xylem vessels that carry water through the plant. So much is produced that the bacteria is actually grown in culture to obtain these polysaccharides, better known as xanthan gum, a thickener used in the food industry.

Some species of Mycoplasma-like species of bacteria also cause serious plant disease. These are bacteria that do not have cell walls and they are introduced into the phloem tissue of plants by insects such as aphids that suck plant sap. Phloem vessels in plants are part of the transport system for sugars, which are generally moved from the leaves, where they are made by photosynthesis, to other parts of the plant, including the roots. Aphids use needle-like mouth parts to puncture plant stems and tap directly into this rich supply of liquid sugar. The Mycoplasma bacteria grow inside the aphids and other sap-sucking insects and are then introduced into their plant host by a very effective insect hypodermic. Plants that produce large amounts of sugar are the worst affected – for example coconut plants and sugar cane. The overall effect of the infection is to weaken the plant, opening it up to other diseases. The leaves usually lose their colour and another common symptom is green flowers, or an absence of flowers.

Plant Bacteria and Humans

Although some of the bacteria that infect plants are from the same large groups as bacterial pathogens in humans, there is no recorded instance of bacteria being passed from plant to person and causing an infection. Plant bacterial diseases are highly specific.

About Bacteria and Antibiotics

What are Bacteria?

Bacteria (singular: bacterium) are single-celled microscopic organisms that often live together in colonies. There are numerous types of bacteria. Most are harmless and some are even helpful, by aiding digestion or breaking down rotting material. A few bacteria can cause diseases in humans, animals and plants; these are referred to as pathogenic bacteria or pathogens.

Where do Bacteria Live?

Excepting normally sterile sites such as the blood stream and spinal fluid, bacteria are commonly found almost anywhere. Their natural habitats are soil, plants, water, and on the skin and in the digestive tracts of humans and animals. They may survive for varying amounts of time on other surfaces that have had contact with these environments.

What is an Antibiotic?

An antibiotic is a chemical compound that can kill or inhibit the growth of a bacteria without causing the host (such as a human or an animal) significant damage. Antibiotics can be naturally produced (like penicillin) by a mold or bacterium, or synthetically made (like the fluoroquinolones).

How do Antibiotics Work?

The activity of antibiotics may be classified as "bacteriostatic" or "bacteriocidal." Bacteriostatic drugs stop the cells from growing and multiplying, but do not actually kill them. This inhibition of growth allows time for the infected host to mount an immune response and eliminate the non-growing infectious agent. Bacteriocidal drugs actually kill the bacteria. These drugs are the preferred choice when a person's immune system is weakened and incapable of destroying the organism itself. In either case, the antibiotics act by disrupting bacterial cell activity, for example, by attacking the outer bacterial cell wall or its inner membrane, or the chemical pathways necessary for the bacteria to live and reproduce. Antibiotics have no activity against viruses.

How are Antibiotics used?

Antibiotics are used in human and veterinary medicine to treat illnesses caused by bacteria and, in some cases, to prevent bacterial infections (prior to surgery, for example). Antibiotics are also used to increase growth in certain food animals, and are used as pesticides to control bacterial infections in food crops.

How Many Antibiotics are There?

There are hundreds of antibiotics, most of which belong to eight major family groups. The drugs are classified according to their basic chemical structure. Most of the members within a family arise from additions or substitutions of attachments to the drug's core structure.

Antibiotic Resistance

What is Antibiotic Resistance?

Antibiotic resistance occurs when an antibiotic has lost its ability to effectively control or kill bacterial growth; in other words, the bacteria are "resistant" and continue to multiply in the presence of therapeutic levels of an antibiotic.

Why do Bacteria become Resistant to Antibiotics?

Antibiotic resistance is a natural phenomenon. When an antibiotic is used, bacteria that can resist that antibiotic have a greater chance of survival than those that are "susceptible." Susceptible bacteria are killed or inhibited by an antibiotic, resulting in a selective pressure for the survival of resistant strains of bacteria. Some resistance occurs without human action, as bacteria can produce and use antibiotics against other bacteria, leading to a low-level of natural selection for resistance to antibiotics. However, the current higher-levels of antibiotic-resistant bacteria are attributed to the overuse and abuse of antibiotics. In some countries and over the Internet, antibiotics can be purchased without a doctor's prescription. Patients sometimes take antibiotics unnecessarily, to treat viral illnesses like the common cold.

How do Bacteria become Resistant?

Some bacteria are naturally resistant to certain types of antibiotics. However, bacteria may also become resistant in two ways: 1) by a genetic mutation or 2) by acquiring resistance from another bacterium. Mutations, rare spontaneous changes of the bacteria's genetic material, are thought to occur in about one in one million to one in ten million cells. Different genetic mutations yield different types of resistance. Some mutations enable the bacteria to produce potent chemicals (enzymes) that inactivate antibiotics, while other mutations eliminate the cell target that the antibiotic attacks. Still others close up the entry ports that allow antibiotics into the cell, and others manufacture pumping mechanisms that export the antibiotic back outside so it never reaches its target.

Bacteria can acquire antibiotic resistance genes from other bacteria in several ways. By undergoing a simple mating process called "conjugation," bacteria can transfer genetic material, including genes encoding resistance to antibiotics (found on plasmids and transposons,) from one bacterium to another. Viruses are another mechanism for passing resistance traits between bacteria. The resistance traits from one bacterium are packaged into the head portion of the virus. The virus then injects the resistance traits into any new bacteria it attacks. Bacteria also have the ability to acquire naked, "free" DNA from their environment. Any bacteria that acquire resistance genes, whether by spontaneous mutation or genetic exchange with other bacteria, have the ability to resist one or more antibiotics. Because bacteria can collect multiple resistance traits over time, they can become resistant to many different families of antibiotics.

How does Antibiotic Resistance Spread?

Genetically, antibiotic resistance spreads through bacteria populations both "vertically," when new generations inherit antibiotic resistance genes, and "horizontally," when bacteria share or exchange sections of genetic material with other bacteria. Horizontal gene transfer can even occur between different bacterial species. Environmentally, antibiotic resistance spreads as bacteria themselves move from place to place; bacteria can travel via airplane, water and wind. People can pass the resistant bacteria to others; for example, by coughing or contact with unwashed hands.

Can Bacteria Lose their Antibiotic Tesistance?

Yes, antibiotic resistance traits can be lost, but this reverse process occurs more slowly. If the selective pressure that is applied by the presence of an antibiotic is removed, the bacterial population can potentially revert to a population of bacteria that responds to antibiotics.

Antibiotic Resistance: A Societal Problem

How can Antibiotic Resistance Harm Humans?

If large numbers of bacteria are resistant to antibiotics, it will be more difficult and more expensive to treat human bacterial infections. When antibiotics fail to work, consequences include extra visits to the doctor, hospitalization or extended hospital stays, a need for more expensive antibiotics to replace the older ineffective ones, lost workdays and, sometimes, death.

Why is Antibiotic Resistance a Public Health Problem?

Antibiotics are called "societal drugs," since antibiotic resistance can pass from bacterium to bacterium, and resistant bacterial infections can pass from person to person. Thus, antibiotic use and antibiotic resistance can eventually affect an entire community.

Why is Antibiotic Resistance an Ecological Problem?

When antibiotics are used in humans or animals, approximately 80-90% of the ingested antibiotics are not broken down, but pass through the body intact and enter the environment as waste. Thus, they retain their ability to affect bacteria and promote antibiotic resistance even after they enter the soil or water as a waste product.

Do People Become resistant to Antibiotics?

No, this is a common misconception. People may exhibit allergic reactions to antibiotics, but they are not resistant to them. It is the bacteria themselves, not the infected host, which become resistant.

How Serious is the Problem of Antibiotic Resistance?

Antibiotic resistance is found all over the world and has become a very serious problem in the treatment of disease. The 1995 US Office of Technology Assessment report attributed a cost of $1.3 billion (1992 dollars) per year for antibiotic-resistant infections due to six species of bacteria in US hospitals. While the real magnitude of the problem is unknown, the monetary cost of treating antibiotic resistant infections worldwide is estimated to be many billions of dollars per year. Some experts predict that, as resistance to antibiotics is increasing at a faster pace than it can be controlled, the future will resemble the preantibiotic era. Others are more optimistic that research and careful drug management can reverse the trend if global efforts are focused on recognizing and controlling it.

Bacteria as Plants

During the last fifteen years the subject of bacteriology has developed with a marvellous rapidity. At the beginning of the ninth decade of the century bacteria were scarcely heard of outside of scientific circles, and very little was known about them even among scientists. Today they are almost household words, and everyone who reads is beginning to recognise that they have important relations to his everyday life. The organisms called bacteria comprise simply a small class of low plants, but this small group has proved to be of such vast importance in its relation to the world in general that its study has little by little crystallized into a science by itself. It is a somewhat anomalous fact that a special branch of science, interesting such a large number of people, should be developed around a small group of low plants.

The importance of bacteriology is not due to any importance bacteria have as plants or as members of the vegetable kingdom, but solely to their powers of producing profound changes in Nature. There is no one family of plants that begins to compare with them in importance. It is the object of this work to point out briefly how much both of good and ill we owe to the life and growth of these microscopic organisms.

As we have learned more and more of them during the last fifty years, it has become more and more evident that this one little class of microscopic plants fills a place in Nature's processes which in some respects balances that filled by the whole of the green plants. Minute as they are, their importance can hardly be overrated, for upon their activities is founded the continued life of the animal and vegetable kingdom. For good and for ill they are agents of neverceasing and almost unlimited powers.

Historical

The study of bacteria practically began with the use of the microscope. It was toward the close of the seventeenth century that the Dutch microscopist, Leeuwenhoek, working with his simple lenses, first saw the organisms which we now know under this name, with sufficient clearness to describe them. Beyond mentioning their existence, however, his observations told little or nothing. Nor can much more be said of the studies which followed during the next one hundred and fifty years. During this long period many a microscope was turned to the observation of these minute organisms, but the majority of observers were contented with simply seeing them, marvelling at their minuteness, and uttering many exclamations of astonishment at the wonders of Nature. A few men of more strictly scientific natures paid some attention to these little organisms. Among them we should perhaps mention Von Gleichen, Muller, Spallanzani, and Needham. Each of these, as well as others, made some contributions to our knowledge of microscopical life, and among other organisms studied those which we now call bacteria. Speculations were even made at these early dates of the possible causal connection of these organisms with diseases, and for a little the medical profession was interested in the suggestion. It was impossible then, however, to obtain any evidence for the truth of this speculation, and it was abandoned as unfounded, and even forgot-ten completely, until revived again about the middle of the 19th century. During this century of wonder a sufficiency of exactness was, how-ever, introduced into the study of microscopic organisms to call for the use of names, and we find Müller using the naines of Monas, Proteus, Vibrio, Bacillus, and Spirillum, names which still continue in use, although commonly with a different significance from that given them by Müller. Müller did indeed make a study sufficient to recognise the several distinct types, and attempted to classify these bodies. They were not regarded as of much importance, but simply as the most minute organisms known.

Nothing of importance came from this work, however, partly because of the inadequacy of the microscopes of the day, and partly because of a failure to understand the real problems at issue. When we remember the minuteness of the bacteria, the impossibility of studying any one of them for more than a few moments at a time—only so long, in fact, as it can be followed under a microscope; when we remember, too, the imperfection of the compound microscopes which made high powers practical impossibilities; and, above all, when we appreciate the looseness of the ideas which pervaded all scientists as to the necessity of accurate observation in distinction from inference, it is not strange that the last century gave us no knowledge of bacteria beyond the mere fact of the existence of some extremely minute organisms in different decaying materials. Nor did the 19th century add much to this until toward its middle. It is true that the microscope was vastly improved early in the century, and since this improvement served as a decided stimulus to the study of microscopic life, among other organisms studied, bacteria received.

Some attention. Ehrenberg, Dujardin, Fuchs, Perty, and others left the impress of their work upon bacteriology even before the middle of the century. It is true that Schwann shrewdly drew conclusions as to the relation of microscopic organisms to various processes of fermentation and decay—conclusions which, although not accepted at the time, have subsequently proved to be correct. It is true that Fuchs made a careful study of the infection of" blue milk," reaching the correct conclusion that the infection was caused by a microscopic organism which he discovered and carefully studied.

It is true that Henle made a general theory as to the relation of such organisms to diseases, and pointed out the logically necessary steps in a demonstration of the causal connection between any organism and a disease. It is true also that a general theory of the production of ail kinds of fermentation by living organisms had been advanced. But all these suggestions made little impression. On the one hand, bacteria were not recognised as a class of organisms by themselves-were not, indeed, distinguished from yeasts or other minute animalculæ.

Their variety was not mistrusted and their significance not conceived. As microscopic organisms, there were no reasons for considering them of any more importance than any other small animals or plants, and their extreme minuteness and simplicity made them of little interest to the microscopist. On the other hand, their causal connection with fermentative and putrefactive processes was entirely obscured by the overshadowing weight of the chemist Lie-big, who believed that fermentations and putrefactions were simply chemical processes. Liebig insisted that all albuminoid bodies were in a state of chemically unstable equilibrium, and if left to themselves would fall to pieces without any need of the action of microscopic organisms.

The force of Liebig's authority and the brilliancy of his expositions led to the wide acceptance of his views and the temporary obscurity of the relation of microscopic organisms to fermentative and putrefactive processes. The objections to Liebig's views were hardly noticed, and the force of the experiments of Schwann was silently ignored. Until the sixth decade of the century, therefore, these organisms, which have since become the basis of a new branch of science, had hardly emerged from obscurity. A few microscopists recognised their existence, just as they did any other group of small animals or plants, but even yet they failed to look upon them as forming a distinct group. A growing number of observations was accumulating, pointing toward a probable causal connection between fermentative and putrefactive processes and the growth of microscopic organisms; but these observations were known only to a few, and were ignored by the majority of scientists.

It was Louis Pasteur who brought bacteria to the front, and it was by his labours that these organisms were rescued from the obscurity of scientific publications and made objects of general and crowning interest. It was Pasteur who first successfully combated the chemical theory of fermentation by showing that albuminous matter had no inherent tendency to decomposition. It was Pasteur who first clearly demonstrated that these little bodies, like all larger animals and plants, come into existence only by ordinary methods of reproduction, and not by any spontaneous generation, as had been earlier claimed. It was Pasteur who first proved that such a common phenomenon as the souring of milk was produced by microscopic organisms growing in the milk. It was Pasteur who first succeeded in demonstrating that certain species of microscopic organisms are the cause of certain diseases, and in suggesting successful methods of avoiding them. All these discoveries were made in rapid succession. Within ten years of the time that his name began to be heard in this connection by scientists, the subject had advanced so rapidly that it had become evident that here was a new subject of importance to the scientific world, if not to the public at large. The other important discoveries which Pasteur made it is not our purpose to mention here. His claim to be considered the founder of bacteriology will be recognised from what has already been mentioned. It was not that he first discovered the organisms, or first studied them; it was not that he first suggested their causal connection with fermentation and disease, but it was because he for the first time placed the subject upon a firm foundation by proving with rigid experiment some of the suggestions made by others, and in this way turned the attention of science to the study of microorganisms.

After the importance of the subject had been demonstrated by Pasteur, others turned their attention in the same direction, either for the purpose of verification or refutation of Pasteur's views. The advance was not very rapid, however, since bacteriological experimentation proved to be a subject of extraordinary difficulty. Bacteria were not even yet recognised as a group of organisms distinct enough to be grouped by themselves, but were even by Pasteur at first confounded with yeasts. As a distinct group of organisms they were first distinguished by Hoffman in 1869, since which date the term bacteria, as applying to this special group of organisms, has been coming more and more into use. So difficult were the investigations, that for years there were hardly any investigators besides Pasteur who could successfully handle the subject and reach conclusions which could stand the test of time. For the next thirty years, although investigators and investigations continued to increase, we can find little besides dispute and confusion along this line. The difficulty of obtaining for experiment any one kind of bacteria by itself, unmixed with others (pure cultures), rendered advance almost impossible. So conflicting were the results that the whole subject soon came into almost hopeless confusion, and very few steps were taken upon any sure basis. So difficult were the methods, so contradictory and confusing the results, because of impure cultures, that a student of today who wishes to look up the previous discoveries in almost any line of bacteriology need hardly go back of 1880, since he can almost rest assured that anything done earlier than that was more likely to be erroneous than correct.

The last fifteen years have, however, seen a wonderful change. The difficulties had been mostly those of methods of work, and with the ninth decade of the century these methods were simplified by Robert Koch. This simplification of method for the first time placed this line of investigation within the reach of scientists who did not have the genius of Pasteur. It was now possible to get pure cultures easily, and to obtain with such pure cultures results which were uniform and simple. It was now possible to take steps which had the stamp of accuracy upon them, and which further experiment did not disprove. From the time when these methods were thus made manageable the study of bacteria increased with a rapidity which has been fairly startling, and the information which has accumulated is almost formidable. The very rapidity with which the investigations have progressed has brought considerable confusion, from the fact that the new discoveries have not had time to be properly assimilated into knowledge. Today many facts are known whose significance is still uncertain, and a clear logical discussion of the facts of modern bacteriology is not possible. But sufficient knowledge has been accumulated and digested to show us at least the direction along which bacteriological advance is tending, and it is to the pointing out of these directions that the following pages will be devoted.

Industrial Enzymes Derived from Plants, Vegetables, Fungus and Bacteria

Enzymes

Alpha Acetolactate De-carboxilase: Alpha-Acetolactate decarboxylase is used to avoid formation of the unpleasant tasting alpha-diacetyl from alpha-acetolactate during fermentation. In the traditional brewing process, the alpha-diacetyl formed from alpha-acetolactate is further reduced to acetoin over a 2-to 4-week maturation period alpha-Acetolactate decarboxylase causes direct decarboxylation of alpha-acetolactate to acetoin, thus avoiding the need for this maturation period. The enzyme can similarly be used in the fermentation of alcohol, where diacetyl is otherwise formed and requires a maturation step before distillation.

Amyloglucosidase

Amyloglucosidase (EC 3.2.1.3) is an important industrial enzyme that is used by the manufacturers of high fructose corn syrup (HFCS). More than half of all commercial baked goods and practically all soft drink bottlers use this syrup instead of sugar because HFCS is both sweeter and cheaper. Amyloglucosidase catalyzes the breakdown of malto-oligosaccharides to glucose.

Bacterial Alpha Amylase

Amylase is a digestive enzyme classified as a saccharidase (an enzyme that cleaves polysaccharides). It is mainly a constituent of pancreatic juice and saliva, needed for the breakdown of long-chain carbohydrates (such as starch) into smaller units. Amylase is a digestive enzyme made primarily by the pancreas and salivary glands. Enzymes are substances made and used by the body to trigger specific chemical reactions. The primary function of the enzyme amylase is to break down starches in food so that they can be used by the body.

Barley Beta-Amylase

Beta-amylase is a 1,4-alpha-D-glucan maltohydrolase (I.U.B. 3.2.1.2) and is present in ungerminated and germinated (malted) barley. Beta-amylase occurs as several isozymes which are immunologically identical. Beta-amylase in ungerminated barley is less soluble in water and migrates more rapidly at pH 8.2 than that of malted barley, but that the two are antigenically identical.

Beta-Glucanase

The glucanase will attack on the cellulose fibre to liberate smaller fragments of cellulose which is further attacked by exo-cellulase to liberate glucose.

Catalase

The ability of organisms to use molecular oxygen was a major evolutionary breakthrough that enabled the production of significantly more energy from the breakdown of foods, amongst many other advantages. However, these advantages came at a cost: toxic by-products known as reactive oxygen species (ROS) are produced, which if left unchecked would seriously affect an organism's viability. These ROS include hydrogen peroxide, superoxide anion radicals, singlet oxygen, hydroxyl radicals and nitric oxide.

Cellulase

Cellulase is an enzyme complex which breaks down cellulose to beta-glucose. It is produced mainly by symbiotic bacteria in the ruminating chambers of herbivores. Aside from ruminants, most animals (including humans) do not produce cellulase, and are therefore unable to use most of the energy contained in plant material

Cellulase refers to a family of enzymes which act in concert to hydrolyze cellulose. Cellulases are widely distributed throughout the biosphere and are most manifest in fungal and microbial organisms.

Coenzyme Q10

Coenzyme Q10 (CoQ 10) or ubiquinone is essentially a vitamin or vitamin-like substance. Disagreements on nomenclature notwithstanding, vitamins are defined as organic compounds essential in minute amounts for normal body function acting as coenzymes or precursors to coenzymes. They are present naturally in foods and sometimes are also synthesized in the body. CoQ10 likewise is found in small amounts in a wide variety of foods and is synthesized in all tissues.

Dextranase

Dextran is a chemically and physically complex polymer, breakdown of which is carried out by a variety of endo-and exodextranases. Enzymes in many groups can be classified as dextranases according to function: such enzymes include dextranhydrolases, glucodextranases, exoisomaltohydrolases, exoisomaltotriohydrases, and branched-dextran exo-1,2--glucosidases. Cycloisomalto-oligosaccharide glucanotransferase does not formally belong to the dextranases even though its side reaction produces hydrolyzed dextrans. A new classification system for glycosylhydrolases and glycosyltransferases, which is based on amino acid sequence similarities, divides the dextranases into five families.

Fungal Alpha Amylase

The enzyme a-amylase from Aspergillus oryzae used in bakeries to improve the bread quality has been identified as an inhalative allergen in baker's asthma. It is doubtful whether this enzyme can induce allergic sensitization in regular bread consumers. The monitoring of the fungal and wheat alpha-amylase activity is of great importance to grain elevators, grain inspectors, flour mills, wheat growers, fungal alpha-amylase manufacturers, and bakeries.

Produced by Aspergillus oryzae and Aspergillus niger mold, fungal alpha-amylase is added to flour by bakers to increase the quality of baked goods. Wheat alpha-amylase naturally occurs in wheat that is exposed to excessive moisture in the fields. Bakers and millers need to quickly and easily determine the exact level of wheat alpha-amylase in flour and fungal alpha-amylase in the additive.

Enzymes I

Fungal Diastase

The Fungal Diastase Or Fungal Alpha Amylase is a potent, dependable, powerful enzyme complex, obtained from Aspergillus oryzae. The Fungal Diastase is manufactured by controlled fermentation of Fungi Aspergillus oryzae, followed by a sophisticated down stream process of recovery.

Fungal Lactase

Lactase is an enzyme in the small intestine that digests lactose (the naturally occurring sugar in milk). Normally lactase breaks lactose down into a form that can be easily absorbed by the blood. Milk, ice cream, yogurt and other dairy products have varying levels of lactose, which affect how much lactase is required for proper digestion.

Fungal Lipase

Fungal Lipase is manufactured from genus aspergillus by fermentation. This is followed by sophisticated down stream processes, which includes filtration, concentration, micro filtration and finally spray drying to fine powder.

Glucoamylase

Glucoamylase (EC 3.2.1.3) is an important industrial enzyme that is used by the manufacturers of high fructose corn syrup (HFCS). More than half of all commercial baked goods and practically all soft drink bottlers use this syrup instead of sugar because HFCS is both sweeter and cheaper. Glucoamylase catalyzes the breakdown of malto-oligosaccharides to glucose.

Amylase is a digestive enzyme classified as a saccharidase (an enzyme that cleaves polysaccharides). It is mainly a constituent of pancreatic juice and saliva, needed for the breakdown of long-chain carbohydrates (such as starch) into smaller units.

Glucose Oxidase

Glucose oxidase (S-?D-glucose: oxygen 1-oxidoreductase, EC1.1.3.4) catalyses the oxidation of S-D-glucose to D-glucono-1,5-lactone and hydrogen peroxide, using molecular oxygen as the electron acceptor.

Glucose oxidase is widely used for the determination of glucose in body fluids and in removing residual glucose and oxygen from beverages and foodstuffs. Furthermore, Glucose oxidase-producing moulds such as Aspergillus and Penicillum Species are used for the biological production of gluconic acid.

Hemicellulase

Hemicellulase is an enzyme that breaks down hemicellulose. Hemicellulose catagorizes a variety of polysaccharides that are more complex than sugars and less complex than cellulose, that are found in plant walls.

It is a mixture of enzymes which can hydrolyze the indigestible components of plant fibres. Since humans lack the endogenous enzymes required to digest plant fibres, the supplementation of hemicellulase provides humans with an additional source of nutrition and reduces the bulking effect of fibrous foods. Scientific evidence suggests that carbohydrolytic enzymes, such as Hemicellulase, can be useful supplements for digestive support and general nutritional support.

Invertase

Invertase is a yeast-derived enzyme. Invertase splits sucrose into glucose and fructose (invert syrup) and can be applied for any inversion of sucrose especially liquefied cherry centres, creams, mints, truffles, marshmallow, invert syrup and other fondants. Invertase is used to improve shelf life of confections.

The official name for invertase is beta-fructofuranosidase (EC3.2.1.26), which implies that the reaction catalyzed by this enzyme is the hydrolysis of the terminal nonreducing beta-fructofuranoside residues in beta-fructofuranosides. Note that alpha-D-glucosidase, which splits off a terminal glucose unit, can also catalyze this reaction. Note that sucrose can be hydrolyzed relatively easily; the reaction proceeds in an acidic environment without the aid of invertase.

Laccase

Laccase is a copper-containing enzyme, 1,4-benzenediol oxidase (EC 1.10.3.2), found in higher plants and microorganisms. Laccases are multicopper oxidases of wide specificity that carry out one-electron oxidation of phenolic and related compounds, and reduce O2 to water. The enzymes are polymeric and generally contain one each of type 1, type 2, type 3 copper centres per subunit, where the type 2 and type 3 are close together forming a trinuclear copper cluster.

Malt Diastase

Malt diastase is a carbohydrolytic enzyme useful for digestive support and general nutrition support. Malt diastase is characterized by the ability to break down amylose and other polysaccharides. The enzyme works with amylase and glucoamylase to digest carbohydrate rich foods such as grains as well as malt, maltose, and sugars. Malt diastase is also known as maltase. It is produced by the cells lining the small intestine.

Nattokinase

Nattokinase is a potent fibrinolytic enzyme extracted and highly purified from a traditional Japanese food called Natto. Natto is a fermented cheese-like food that has been used in Japan for over 1000 years for its popular taste and as a folk remedy for heart and vascular diseases.

Natto is produced by a fermentation process by adding Bacillus natto, a beneficial bacteria, to boiled soybeans. The resulting Nattokinase enzyme, is produced when Bacillus natto acts on the soybeans. While other soy foods contain enzymes, it is only the natto preparation that contains the specific nattokinase enzyme.

Neutral Cellulase

The activity of any enzyme depends primarily on the pH and temperature of the reaction system. Basically, cellulases can be divided in to four basic groups according to the pH required for optimum enzymatic activity.

One group is called Acid Cellulases because the pH ranges from 4.5 to 5.0. There are Hybrid Cellulases whose optimum pH is 4.5 to 7.0. The third group called the Neutral Cellulases is active from a pH 6.0 to 8.0 but is optimal at pH 6.2. The fourth group is commonly referred as to Alkaline Cellulase enzymes because its optimum pH is 7.2 to 8.5.

Enzymes II

Ox Bile Extract

Bile is a secretion of liver, stored in the gall bladder until needed. Bile has a major role to play in fat digestion and absorption. The active constituents of bile are the bile salts, which are either glycine or taurine conjugates of polyhydroxy steroidal acid. Principal bile acids of mammals are the hydroxyl derivatives of 5-beta-cholan-24 oic-acid. Ox-bile is the most important commercial source of these acids and contains primarily Cholic acid with less amount of deoxycholic acid.

Pancreatin

Pancreatin is a digestive enzyme that is used to supplement loss of or low levels of digestive enzymes, and is often used in people with cystic fibrosis. It has been claimed to help with food allergies, celiac disease, autoimmune disease, cancer, and weight loss. Pancreatin is also known as pancreatic acid.

Papain

Papain is naturally occurring enzyme, obtained from latex of fruit of PAPAYA (Carica papaya Latex). The enzyme Papain is obtained by drying papaya latex by suitable method viz; sun drying, tray drying, spray drying etc. spray drying gives a fine powder, which dissolves easily in water to provide a solution of high proteolytic activity.

Pectinase

Pectinase is a general term for enzymes that break down pectin, a polysaccharide substrate that is found in the cell walls of plants. One of the must studied and widely used commercial pectinases is polygalacturonase. It is useful because pectin is the jelly-like matrix which helps cement plant cells together and in which other cell wall components, such as cellulose fibrils, are embedded. Therefore pectinase enzymes are commonly used in processes involving the degradation of plant materials, such as speeding up the extraction of fruit juice from fruit, including apples. Pectinases have also been used in wine production since the 1960s.

Pepsin

Pepsin is one of the principal protein degrading or proteolytic enzymes in the digestive system. During the process of digestion, Pepsin acts on the complex dietary protein and breaks up into peptides and amino acids which can be readily absorbed by the intestinal lining.

Peroxidase

Peroxidases are a group of enzymes that catalyze oxidation-reduction reactions. As such, they are classified as oxidoreductases. Toxic molecules such as superoxide and hydroxide radicals can be found in cells due to the presence of oxygen. These are byproducts of aerobic respiration. They are eliminated by a number of enzymes present inside the cell. Superoxide, for example, is destroyed by superoxide dismutase. The degradation, however, produces more hydrogen peroxide (H2O2), which is, in turn, destroyed by peroxidase. Peroxidases reduce H2O2 to water while oxidizing a variety of substrates. Thus, peroxidases are oxidoreductases which use H2O2 as electron acceptor for catalyzing different oxidative reactions.

Phytase

Large amount of research show that phytate or phytic acid in monogastric animal's feeds has a strong anti-nutrition effect the reason being: Monogastric animals lack phytate degrading enzyme in their digestive tract so they are unable to utilize the phosphorus of phytate

Protease

Proteases (Proteinases, Peptidases or Proteolytic enzymes) are enzymes that break peptide bonds between amino acids of proteins. The process is called proteolytic cleavage, a common mechanism of activation or inactivation of enzymes especially involved in blood coagulation or digestion. They use a molecule of water for this and are thus classified as hydrolases.

Pullulanase

Pullulanase is a specific kind of glucanase, an amylolytic exoenzyme, that degrades pullulan. It is produced as an extracellular, cell surface-anchored lipoprotein by Gram-negative bacteria of the genus Klebsiella. Type I pullulanases specifically attack a-1,6 linkages, while type II pullulanases are also able to hydrolyse a-1,4 linkages. It is also produced by some other bacteria and archaea. Pullanase is used as a detergent in biotechnology.

Serratiopeptidase

Serratiopeptidase is derived from bacteria belonging to the genus Serratia. The controlled fermentation of Serratia sp. secretes this enzyme in the highly selective medium. The recovery process involves various types of filtration, concentration and steps to make enzyme useful for pharmaceutical applications and finally dried to fine free flowing powder form.

Sucrase

Sucrase is a yeast-derived enzyme. Sucrase splits sucrose into glucose and fructose (invert syrup) and can be applied for any inversion of sucrose especially liquefied cherry centres, creams, mints, truffles, marshmallow, invert syrup and other fondants. Sucrase is used to improve shelf life of confections. It is available in single, double and triple strengths and is packaged in one, ten and 44 pound containers for ease of use, storage and cost efficiency.

Trypsin Chymotrypsin Mix

Trypsin Chymotrypsin Mix is a unique anti-inflammatory product manufactured by Advanced Enzyme Technologies. Trypsin Chymotrypsin Mix is a proteolytic enzyme obtained by controlled proteolysis of zymogens form in the frozen glands. Successive conc. and purification steps give Trypsin Chymotrypsin Mix best suitable for pharmaceutical applications.

Xylanase

Xylanase produced by submerged fermentation of Trichoderma reesei var. It has cellulase side activity. This is classified by Nomenclature Committee of the International Union of Biochemistry and Molecular Biology (NC-IUBMB) in consultation with the IUPAC-IUBMB joint Commission on Biochemical Nomenclature (JCBN).

5

Nutrition and Growth of Bacteria

Nutritional Requirements of Cells

Every organism must find in its environment all of the substances required for energy generation and cellular biosynthesis. The chemicals and elements of this environment that are utilized for bacterial growth are referred to as nutrients or nutritional requirements.

Many bacteria can be grown the laboratory in culture media which are designed to provide all the essential nutrients in solution for bacterial growth.

Bacteria that are symbionts or obligate intracellular parasites of other cells, usually eucaryotic cells, are (not unexpe-ctedly) difficult to grow outside of their natural host cells.

Whether the microbe is a mutualist or parasite, the host cell must ultimately provide the nutritional requirements of its resident.

Many bacteria can be identified in the environment by inspection or using genetic techniques, but attempts to isolate and grow them in artificial culture has been unsuccessful. This, in part, is the basis of the estimate that we may know less than one percent of all procaryotes that exist.

The Major Elements

At an elementary level, the nutritional requirements of a bacterium such as *E. coli* are revealed by the cell's elemental composition, which consi t s of C, H, O, N, S. P, K, Mg, Fe, Ca, Mn, and traces of Zn, Co, Cu, and Mo. These elements are found in the form of water, inorganic ions, small molecules, and macromolecules which serve either a structural or functional role in the cells.

Trace Elements

Trace elements are metal ions required by certain cells in such small amounts that it is difficult to detect (measure) them, and it is not necessary to add them to culture media as nutrients. Trace elements are required in such small amounts that they are present as "contaminants" of the water or other media components. As metal ions, the trace elements usually act as cofactors for essential enzymatic reactions in the cell. One organism's trace element may be another's required element and vice-versa, but the usual cations that qualify as trace elements in bacterial nutrition are Mn, Co, Zn, Cu, and Mo.

Carbon and Energy Sources for Bacterial Growth

In order to grow in nature or in the laboratory, a bacterium must have an energy source, a source of carbon and other required nutrients, and a permissive range of physical conditions such as O_2 concentration, temperature, and pH. Sometimes bacteria are referred to as individuals or groups based on their patterns of growth under various chemical (nutritional) or physical conditions.

For example, phototrophs are organisms that use light as an energy source; anaerobes are organisms that grow without oxygen; thermophiles are organisms that grow at high temperatures. All living organisms require a source of energy. Organisms that use radiant energy (light) are called phototrophs. Organisms that use (oxidize) an organic form of carbon are called heterotrophs or (chemo)heterotrophs. Organisms that oxidize inorganic compounds are called lithotrophs. The carbon requirements of organisms must be met by organic carbon (a chemical compound with a carbon-hydrogen bond) or by CO_2. Organisms that use organic carbon are heterotrophs and organisms that use CO_2 as a sole source of carbon for growth are called autotrophs.

Almost all eucaryotes are either photoautotrophic (e.g. plants and algae) or heterotrophic (e.g. animals, protozoa, fungi). Lithotrophy is unique to procaryotes and photoheterotrophy, common in the Purple and Green Bacteria, occurs only in a very few eucaryotic algae. Phototrophy has not been found in the Archaea, except for nonphotosynthetic light-driven ATP synthesis in the extreme halophiles.

The Cultivation of the Parasitic Protozoa Without Bacteria

A review of the literature shows that isolation of protozoa from bacteria does not present any formidable difficulties. Such isolation has been accomplished by methods based on principles involving positive or negative geotropism of protozoa, migration of protozoa away from bacteria through semi-solid media, mechanical separation of protozoa and bacteria by micro-isolation, or the differential action of poisons on cysts of protozoa and certain non-spore-forming bacteria. In cultivating isolated protozoa, provision must be made for a gradual adaptation to new conditions.

Physical and chemical considerations which must be taken into account to provide a favourable medium include isotonicity of the fluid part of the medium with those of the tracts parasitized by the protozoa, a balanced food supply and a balanced interaction of various respiratory factors, vitamins and catalysts.

The L.E.S. medium, providing both solid and liquid food, is probably the best for general use that has thus far been devised. Eutrichomastix colubrorum, Trichomonas foetus and T. Columbae, which may be considered as representative flagellate parasites of the digestive, urogenital and respiratory tracts, respectively, all find adequate nourishment in this medium in the absence of bacteria. However, for the growth of Endamoeba histolytica without bacteria the medium lacks certain essential factors, the discovery of which awaits future investigations.

Bacteria help regulate rates of organic matter mineralization, nutrient cycling, and energy transfer in aquatic environments. These ecosystem processes are commonly limited by the quantity and quality of resources, such as dissolved organic matter (DOM). One acknowledged, but possibly underappreciated, source of DOM in aquatic ecosystems is DNA. Dissolved DNA (dDNA) is produced by cell death, lysis, and excretion and is an important component of the DOM pool in marine environments. For example, water column concentrations of dDNA often exceed particulate DNA concentrations, while extracellular DNA in marine sediments represents the largest reservoir of DNA in the world ocean.

The turnover rates for dDNA are rapid (<1 day), suggesting that it may be a high-quality resource capable of supporting microbial metabolism. dDNA helps bacteria meet their carbon and nitrogen demands and is a source of nucleotides that can be directly used for

DNA and RNA synthesis. Due to its elemental composition, dDNA may be an especially valuable source of phosphorus. For example, dDNA accounts for up to 25% of the total phosphorus in some freshwater ecosystems, and it is estimated to provide one-half of the phosphorus required for benthic bacterial biomass production in some marine ecosystems.

The nutritional importance of dDNA in aquatic ecosystems ultimately depends on the ability of microorganisms to take up this macromolecule. It is generally accepted that bacteria are constrained by their inability to passively transport materials with >0.6 kDa molecular masses across their cell walls. Environmental dDNA is much larger than this; its molecular mass ranges from 6 to 23,000 kDa. Therefore, bacteria have evolved different strategies for taking up dDNA. Some marine bacteria are capable of degrading dDNA with extracellular nucleases and then consume the hydrolyzed products.

Alternatively, some bacteria are naturally competent and take up intact dDNA for nutritional and genetic purposes via complex pathways consisting of specialized proteins. In aquatic ecosystems, the size of DOM has a strong effect on the diversity and metabolism of aquatic microorganisms. For example, some groups of marine bacteria are specialized at using different types of low-molecular-weight (LMW) DOM, and manipulation of DOM size fractions has been shown to alter the composition of estuarine bacterial communities.

Interactions between dDNA size and metabolic strategies of different bacteria, therefore, may ultimately influence the rates of dDNA turnover and the patterns of microbial diversity in marine ecosystems.

Despite the potential importance of dDNA as a microbial resource capable of influencing ecosystem processes, very little is known about the diversity, phylogenetic identity, or interactions of heterotrophic bacteria that consume dDNA. In this study, experiments with water from Eel Pond (Woods Hole, MA) were conducted to characterize the diversity of cultivable marine bacteria capable of using dDNA as the primary source of carbon, nitrogen, and phosphorus.

Furthermore, I tested the hypothesis that dDNA size influences bacterial metabolism by comparing the abundance, growth, and nuclease production of bacteria exposed to LMW dDNA and high-molecular-weight (HMW) dDNA. Understanding the ecological strategies of marine bacteria involved in dDNA degradation should provide new insight into microbial species interactions and the factors influencing the biogeochemical processing of dDNA in aquatic ecosystems.

Materials and Methods

Field sampling

Water samples were obtained from Eel Pond, a small impoundment open to Great Harbor in Woods Hole, MA. A peristaltic pump was used to collect water from three depths (0.5, 2.5, and 5.0 m) at a sampling location with a maximum depth of 5.5 m. Water samples were pooled and transported to the laboratory immediately. In the field, temperature, salinity, dissolved oxygen content, and conductivity were measured with a YSI multiprobe water quality checker. The abundance of bacteria in the pooled water samples was determined by 4',6'-diamidino-2-phenylindole (DAPI) staining and epifluorescence microscopy. The dDNA concentration in the pooled sample was determined spectrofluorimetrically using SYBR green I and standard solutions of *Escherichia coli* DNA (Sigma).

Enrichments

Dilutions of the pooled Eel Pond sample (100 μl of 10^0 to 10^{-4} dilutions) were plated onto replicate agar plates containing HMW dDNA, LMW dDNA, or no dDNA (control). The plates were incubated at 30°C in the dark and monitored daily for colony formation. After 6 days, a subset of replicate plates was used for plate wash PCR. After 10 days, bacterial abundance was estimated by calculating the total number of CFU for a set of replicate plates for each of the three dDNA treatments. One-way analysis of variance (ANOVA) and Tukey's HSD (honestly significant difference) were used to test for differences in CFU among DNA treatments.

Diversity of dDNA-consuming Bacteria

Clone libraries for both HMW and LMW dDNA enrichments were constructed by extracting genomic DNA from agar plates using a plate wash technique and an Ultraclean fecal DNA isolation kit (MoBio Laboratories). 16S rRNA genes were amplified using universal bacterial primers Of and 1492r, and the resulting PCR products were cloned using an Invitrogen TOPO TA cloning kit.

A total of 106 clones were picked for sequencing (40 LMW dDNA clones and 66 HMW dDNA clones). Purified plasmids were used as templates for partial sequence determination using bacterial 16S rRNA gene primer 531R (5'-TAC CGC GGC TGC TGG CAC-3') with a BigDye Termination kit, version 3.0 (Applied Biosystems). Completed-reaction mixtures were examined with a 96-capillary 3730xl DNA analyzer (Applied Biosystems).

Sequences were aligned using the ARB phylogenetic software package. Sequences were checked for chimeras using Bellerophon. DOTUR was used to estimate taxon richness, assuming a 99% sequence similarity cutoff. Finally,-LIBSHUFF was used to test the null hypothesis that the HMW and LMW dDNA clone libraries came from a common population.

Phylogenetic Identity of Isolates

Single-colony isolates from HMW and LMW dDNA plates were identified by direct sequencing of 16S rRNA genes. Isolates were randomly selected from the enrichments and replated onto the original dDNA medium. Individual colonies were lysed by boiling them in 20 μl of 0.05% Triton X-100 for 5 min. One microliter of each resulting supernatant was used as a template in a 25-μl PCR mixture. The 37 PCR products were screened by agarose gel electrophoresis and sequenced after they were cleaned with an ExoSAP-IT kit (USB Corporation). Phylogenetic identification of the sequences was performed by determining the most closely related representatives in the BLAST database. A total of 37 isolates were identified (18 isolates from HMW dDNA plates and 19 isolates from LMW dDNA plates).

Growth curves with dDNA

The 19 phylogenetically identified isolates from LMW dDNA plates were used in a set of growth curve experiments performed with liquid dDNA media. These experiments were performed in order to characterize bacterial growth on different sizes of dDNA. The liquid media used were the media described above without agar. For each isolate, there were six test tubes (15 ml); three replicate tubes contained LMW dDNA, and three replicate tubes contained HMW dDNA. The tubes were inoculated with a homogeneous suspension of cells (100 μl of a 0.9% NaCl suspension) and incubated in an upright shaker at 30°C. Growth was monitored by determining the change in absorbance at 600 nm over time (110 h). The effect of dDNA type (HMW dDNA versus LMW dDNA) on bacterial growth over time was assessed using repeated-measures ANOVA (SAS PROC MIXED with an AR-1 covariance matrix). A significant effect of time in the repeated-measures

ANOVA model indicated that growth changed during the experiment irrespective of the dDNA treatment. A significant effect of dDNA indicated that growth with LMW dDNA and growth with HMW dDNA were different irrespective of time. A significant interaction (time x dDNA) indicated that changes in growth over the course of the experiment were affected by the dDNA treatment.

Field Characteristics

The salinity, temperature, and conductivity were uniform throughout the water column in Eel Pond. In contrast, the dissolved oxygen concentration decreased with depth from 7.3 to 3.6 mg liter^{-1}. The concentration of DAPI-stained bacteria in the pooled water sample was 1.3 x $10^7 \pm 0.71$ x 10^7 cells ml^{-1} (average ± standard error). The dDNA concentration in the pooled water sample was 26 ± 0.3 μg liter^{-1} (mean ± standard error).

Diversity of dDNA-consuming Bacteria

Using the 106 sequences generated from the plate wash PCR, 32 taxa were identified with a 99% sequence similarity cutoff. The taxon richness of the LMW dDNA clone library and the taxon richness of the HMW dDNA clone library were comparable (18 and 21 taxa, respectively). However, only 7 of the 32 taxonomic groups (22%) contained sequences obtained from treatments; the remaining 25 taxonomic groups (78%) contained sequences obtained with either LMW dDNA or HMW dDNA.-LIBSHUFF analysis indicated that the compositions of the clone libraries were significantly different (*P* = 0.0001). Most (88%) of the sequences belonged to the-proteobacteria and were represented by *Vibrio* spp., *Alteromonas* spp., and *Pseudoalteromonas* spp. Two sequences belonged to the-proteobacteria; one of these sequences was closely related to a *Roseobacter* sp. (99% similarity), and the other was most closely related to *Kordiimonas gwangyangensis* (90% similarity). In addition, two sequences belonged to the *Bacteroidetes*; one of these sequences was closely related to a *Flexibacter* sp. (95% similarity), and the other was most closely related to *Microscilla arenaria* (94% similarity). Some sequences formed monophyletic groups with respect to the type of dDNA. For example, *Vibrio* sequences were recovered only when LMW dDNA was used. In addition, there were three HMW *Pseudoalteromonas* clades and a single LMW *Alteromonas* clade. The largest clade was the "mixed" *Alteromonas* group, which comprised 17 sequences obtained with LMW dDNA and 29 sequences obtained with HMW dDNA.

Bacterial Abundance with LMW and HMW dDNA

The average number of CFU was higher on LMW dDNA agar plates than on HMW dDNA agar plates ($5.2 \times 10^5 \pm 0.32 \times 10^5$ cells ml^{-1} versus $2.1 \times 10^5 \pm 0.43 \times 10^5$ cells ml^{-1}; $P < 0.0001$, as determined by ANOVA). There were significantly fewer CFU on control plates ($1.5 \times 10^4 \pm 0.99 \times 10^4$ cells ml^{-1}) than on both LMW dDNA and HMW dDNA agar plates ($P < 0.0001$, as determined by ANOVA).

Growth Curves with dDNA

A majority (58%) of the bacteria that were isolated from the LMW dDNA enrichments were affected by the type of dDNA in the growth curve experiments. In all but one instance (*Pseudoalteromonas* isolate 5), isolates obtained from the LMW dDNA enrichments had higher levels of growth in the presence of LMW dDNA than in the presence of HMW dDNA. The remaining isolates (42%) exhibited similar growth in the presence of LMW dDNA and HMW dDNA.

Discussion

dDNA is a ubiquitous component of the DOM pool in nearly all aquatic environments. The biogeochemical cycling of dDNA largely depends on microbial activity. dDNA may be a high-quality resource capable of supporting aquatic microbial metabolism, yet very little is known about the identity and physiology of dDNA-consuming bacteria. Some, but not all, of the marine microbes isolated in this study were able to take up and efficiently use dDNA as the primary source of energy and nutrition. Bacterial isolates typically exhibited less growth on HMW dDNA than on LMW dDNA, especially if they were originally isolated from LMW dDNA enrichments. This difference in bacterial growth may reflect energetic costs associated with nuclease production or, in some cases, an inability to produce nucleases altogether. The results of this study suggest that dDNA size can influence microbial diversity, but they also imply that different bacterial species may specialize on certain size fractions of the dDNA pool. Thus, dDNA turnover in marine ecosystems may involve a succession of microbial assemblages with different ecological strategies.

Diversity of dDNA-consuming Bacteria

A relatively diverse assemblage of marine bacteria was cultivated in this study, especially considering that a single molecule was used as the primary source of carbon, nitrogen, and phosphorus. Members of the *Bacteroidetes* and-proteobacteria comprised 12% of the taxa in the clone libraries, while the remainder were members of the-proteobacteria. The latter group was dominated by *Alteromonas* and *Pseudoalteromonas*, which are well known for their ability to degrade HMW compounds via extracellular enzyme production.

However, culture-independent surveys have shown that the abundance of-proteobacteria is generally low compared to other groups of bacteria in marine ecosystems. Despite being relatively rare, the-proteobacteria comprise an ecologically important group of microbes that tend to be fast growing under resource-rich conditions. Marine systems tend to experience episodic pulses of high resource availability due to both large-scale processes (e.g., upwelling, eddies, or bloom dynamics) and small-scale processes (e.g., sloppy grazing and viral lysis). Some studies have suggested that the-proteobacteria are able to take advantage of these resource pulses.

For example, in southern California, the abundance of-proteobacteria increased in response to a dinoflagellate bloom and experimental DOM enrichment. Due to the biological controls on its production, dDNA may be an example of the type of resource that helps opportunistic bacteria, like the-proteobacteria, persist over time in marine environments. It is important to note that dDNA consumption is most likely not restricted to the-proteobacteria or other taxa found in this study. The cultivation approach used here may provide only a glimpse of the true diversity of microbes capable of using dDNA as a source of energy or nutrition in marine environments.

dDNA Size Affects Microbial Composition

dDNA size had a strong effect on the composition of the bacteria represented in the clone libraries. Initially, I predicted that bacteria which grew on HMW dDNA would be a subset of the bacterial assemblage which grew on LMW dDNA. Such an outcome would be consistent with a loss of taxon richness due to the inability of some bacteria to efficiently take up and use HMW dDNA. In contrast to the initial prediction, the taxon richness obtained with LMW dDNA and the taxon richness obtained with HMW dDNA were comparable.

The bacterial composition, however, was significantly affected by dDNA size. Only 22% of the taxa were capable of using both types of dDNA, while the vast majority of bacteria were limited to growing on either LMW or HMW dDNA. Additional effects of dDNA size on microbial composition were apparent from the structure of the phylogenetic tree constructed from the clone libraries. For example, a number of *Pseudoalteromonas* groups were monophyletic with respect to HMW dDNA, while *Vibrio* clones were only recovered from the LMW dDNA treatment. Other studies have also shown that DOM size affects aquatic microbial composition. An estuarine bacterial community from a blackwater river was strongly influenced by exposure to different size fractions of freshwater DOM.

Although the estuarine bacterial community was affected by substrate size, it may have also been responding to variation in the chemical composition of the different DOM size fractions. For example, the composition of the bacterial community after DOM additions in the Delaware Bay was dependent on the type of HMW molecule investigated. In contrast, the only apparent factor in this study that could account for treatment-level differences in microbial composition was the size or structure of the dDNA molecule.

Bacterial growth Affected by dDNA Size

The changes in microbial composition corresponded to trends in the abundance and growth of bacteria on LMW and HMW dDNA. First, bacterial abundance was approximately 2.5 times greater on LMW dDNA plates than on HMW dDNA plates. Second, one-half of the strains isolated from LMW dDNA enrichments exhibited reduced growth when they were inoculated into HMW dDNA liquid medium. Together, these results demonstrate that dDNA metabolism varies among marine bacteria but in general is strongly affected by dDNA size. These findings are consistent with the view that DOM uptake by bacteria is constrained by their ability to take up HMW molecules.

The results of the growth curve experiments also provide insight into a potential mechanism that may have contributed to the observed differences in microbial composition. Specifically, bacteria isolated from the LMW dDNA enrichments generally performed better when they were grown in LMW dDNA liquid medium and thus may have outcompeted HMW strains under these conditions.

DNA Degradation Varies Among Marine Bacteria

The results of the ethidium bromide assay clearly showed that bacteria isolated from different enrichments degraded dDNA at different rates. For example, many of the isolates originating from the LMW dDNA enrichment degraded <1% of the available HMW dDNA, suggesting that they lacked the ability to produce extracellular nucleases. In contrast, bacteria isolated from the HMW dDNA enrichments degraded HMW dDNA two times faster than bacteria isolated from the LMW dDNA enrichments.

The ethidium bromide assay, which has not been used previously to assess dDNA degradation by marine bacteria, is similar to assays used for measuring nuclease activity. Therefore, it is reasonable to assume that the DNA degradation observed in this study reflected some integrated features of bacterial nuclease production, which might include variation in specific activities, production rates, or diffusive properties of different enzymes. Moreover, the results of the DNA degradation assay could be influenced by the physical location of the enzymes (in particular, whether nucleases tend to be bound to the cell surface or released freely into the environment).

Results of this study indicate there may be substantial variation in dDNA degradation even among closely related taxa. For example, *Pseudoalteromonas* spp. degraded DNA at comparable rates regardless of the enrichment conditions. In contrast, the dDNA degradation rates were three times greater for *Alteromonas* spp. isolated from HMW dDNA enrichments than for *Alteromonas* spp. isolated from LMW dDNA enrichments, suggesting that observed "microdiversity" may translate into phenotypic diversity. It is important to note, however, that degradation is only the first step involved in dDNA metabolism. The overall metabolism of dDNA-consuming bacteria should also be influenced by factors such as uptake kinetics and yield, which are reflected in growth dynamics.

Currently, two models are used to explain bacterial uptake of dDNA for nutritional purposes. According to the first model, some bacteria are naturally competent and capable of taking up intact macromolecular DNA. Competency requires proteins related to type IV pili and type II secretion systems, in addition to cell surface ectonuclease production. Competency often results in natural transformation, a mechanism for horizontal gene transfer, which may be important for explaining the evolutionary history of some marine bacteria. Competency has also been suggested to be a potential nutritional strategy for some bacteria but it has not been well documented for marine microbes. One of the few studies on this topic found that estuarine bacteria were ineffective at transporting intact gene sequences from the environment into the cell. Such findings support the second model, whereby bacteria degrade dDNA extracellularly with the aid of nucleases and then consume the hydrolyzed products· Early work on this question suggested that many bacteria isolated from marine environments, including *Pseudomonas* spp., *Flavobacterium* spp., *Vibrio* spp., and *Acinetobacter* spp., are capable of using this DNA consumption strategy.

Although the current study was not designed to differentiate the mechanisms of dDNA uptake in marine bacteria, the results may shed some light on this topic. For example, dDNA degradation assays revealed wide zones of clearing around confined regions of bacterial biomass. This pattern of dDNA uptake is more consistent with extracellular enzyme degradation than with competency, which requires binding of DNA by the bacterial cell· Thus, the results of the current study support the model involving extracellular dDNA degradation followed by uptake of the hydrolyzed products, but additional research is needed to better understand the relative importance of extracellular enzyme hydrolysis and competency strategies in dDNA uptake by marine bacteria.

Ecological Implications

Results obtained in this study indicated that populations of co-occurring marine bacteria were specialized for using either LMW or HMW dDNA. On average, bacteria isolated from LMW dDNA enrichments exhibited greater growth on LMW dDNA than on HMW dDNA, while bacteria isolated from HMW dDNA enrichments were more effective at enzymatic degradation of HMW dDNA than bacteria isolated from LMW dDNA enrichments· This type of resource specialization on different sizes of dDNA suggests that there is a trade-off which could help maintain microbial diversity and influence ecosystem processes. For example, resource specialization by distinct groups of bacteria in a microbial assemblage has been shown to influence rates of DOM consumption in marine ecosystems.

Some features of the dDNA experimental system are similar to features of a recent simulation model that was used to examine the interactions between microbial populations receiving HMW organic substrates. In this model there were two types of bacteria: bacteria that were directly responsible for HMW substrate degradation via extracellular enzymes ("producers") and bacteria that took advantage of the extracellular enzymes generated by producers ("cheaters"). Simulations revealed that cheaters had a selective advantage when the enzyme cost increased for producers. In contrast, producers were favoured in heterogeneous environments or when enzymes remained close to the producer's cell surface. Ultimately, the model predicted that producers and cheaters would coexist at intermediate levels of enzyme cost and intermediate levels of enzyme diffusivity. The results of this experimental study suggest that the dDNA system may be ideal for testing theoretical predictions about extracellular enzyme activity, microbial interactions, and DOM dynamics.

Previous studies have documented that there is rapid uptake of dDNA by aquatic microbial communities, suggesting that dDNA is an important component of DOM in freshwater and marine ecosystems. Despite this, very little is known about the identity and metabolism of environmental bacteria capable of using dDNA as a source of energy or nutrition. Using a novel cultivation approach, this study provides an initial assessment of the diversity of dDNA-consuming bacteria in marine ecosystems.

Almost certainly, the list of dDNA-consuming microbes will grow with the development and application of cultivation-independent approaches. Importantly, the results obtained in this study highlight some of the ecological strategies that marine bacteria use for consuming dDNA. Moreover, the results suggest that the biogeochemical cycling of dDNA in marine ecosystems may involve a succession of microbial assemblages with different metabolic capabilities.

A Membrane-lid Tube for Bacteria Cultivation and Freeze-drying

The membrane-lid tube Lid_{Bac} from Eppendorf is a significant contributory factor in the miniaturization process to the microbiological laboratory. For the first time ever, it is now possible to cultivate miniprep bacteria cultures in exactly the format required for the preparation. Following cultivation, the bacteria no longer have to be pipetted into a new tube since the culture is grown in the tube of the subsequent preparation.

The automated production process enables the membrane-lid tube to be used directly, without need for prior sterilization. The cleaning process before and after cultivation, as required when conventional glass tubes are used, is no longer necessary. Target The experiments described in this article demonstrate the suitability of the membrane-lid tube Lid_{Bac} for bacteria cultivation on a miniprep scale (1 to 2 ml). A comparison is made between the growth behaviour of bacteria in Lid_{Bac} and in conventional Erlenmeyer flasks.

The membrane-lid tube is suitable for virtually all stages of the experiment whenever ventilation, drying, or a similar step is necessary.

This is underlined by a comparison of the results obtained by freeze-drying in Lid_{Bac} with those produced in a standard tube.

Material and Methods Lid_{Bac}

Lid_{Bac} consists of a polypropylene (PP) membrane-lid with an integrated PP hydrophobic membrane and a Safe-Lock microcentrifuge tube. The membrane is gas-permeable, hydrophobic,and impermeable to bacteria (0.2 m pore size). The membrane lid is compatible with all 1.5 ml and 2 ml Eppendorf Tubes. After the tube has been filled, the membrane-lid is simply pressed into the tube. As there is no hinged connection to the tube, the lid can be removed easily at any time to enable the tube to be used for follow-up applications. Bacteria cultivation Growth kinetics: Four parallel experiments were carried out in order to investigate the growth behaviour of Escherichia coli HB101 and Bacillus subtilis 168 when cultivated in Lid_{Bac}. A positive control was carried out in parallel in the form of bacteria cultivation in Erlenmeyer flasks. To ensure an identical initial bacteria concentration in the different test tubes, LB (Luria Bertani) medium was given a 1% inoculation of overnight culture. After being mixed thoroughly, 1 ml bacteria culture each was transferred to 1.5 ml membrane-lid tubes (Eppendorf Lid_{Bac})and 5 ml each was transferred to a 50 ml Erlenmeyer flask with cellulose stopper.

The membrane-lid tubes were then incubated in the Eppendorf Thermomixer R at 1,400 rpm and 37C. The Erlenmeyer flasks were sealed with cellulose stoppers and incubated in a linear oscillating water bath. The growth kinetics of the bacteria were determined by means of absorption measurements at 620 nm.

Evaporation: Evaporation experiments were carried out with water since no significant differences were expected from experiments with culture medium containing growth. 1 ml water was transferred into each 1.5 ml Lid_{Bac} tube. The tubes were sealed with their membrane-lid and then incubated in the Thermomixer R at 37 C and 1,400 rpm (room temperature 22-23 C, relative humidity 45-50 %). As a positive control, 5 ml water was transferred into each 50 ml Erlenmeyer flask, which was then sealed with a cellulose stopper and incubated in a linear oscillating water bath at 37 C. The weight of the tubes was determined in five parallel measurements, carried out after 3, 6, 9,18, and 24 hours.

Cell morphology: Since the cell morphology of bacteria often changes when insufficient amounts of oxygen are supplied for example, under sub-optimal air conditions, long, thin nutrient-deprived phenotypes are formed in bacteria that normally grow in short, rodlike shapes an analysis of cell morphology of cultivated bacteria enables the tubes to be assessed in regard to their suitability for bacteria cultivation.

Following overnight cultivation, the bacteria strains E. coli HB101 (facultatively anaerobic) and B. subtilis 168 (strictly aerobic) were inoculated at a concentration of 1% into the LB medium. 1 ml of this medium was then transferred into each 1.5 ml Lid_{Bac} tube and 5 ml of the medium was placed into each 50 ml Erlenmeyer flask.

The vessels were then sealed (Lid_{Bac} with its membrane-lid, the Erlenmeyer flasks with cellulose stoppers) and incubated under the same conditions as for bacteria cultivation. After 3, 6,16, and 24 hours, samples were removed from both vessel types and examined under a microscope. To prevent any effect on the morphology of the bacteria caused by the increase in the oxygen supply as a result of the tubes being opened, a new tube was used for each measurement. Freeze-drying In five parallel preparations, a comparison was made between freeze-drying E. coli HB101 in membrane-lid tubes and in standard tubes.

Tests for Identification of Bacteria

Most of these are procedures taught in introductory microbiology courses, and are useful in sorting bacteria into broad categories rather than identifying bacteria to the species level.

Differential Stains

Differential bacterial stains are the application of a series of dyes that result in some organisms turning one colour while other physically or chemically distinct bacteria retain a contrasting colour. Gram stain, Acid Fast-stain and Endospore stain are differential staining procedures used to classify bacteria into general groups. Gram stain distinguishes Gram+ from Gram– bacteria based on differences in bacterial cell wall structure. The Acid-fast stain protocol helps differentiate bacteria that have a waxy cell wall from those that do not. The Endospore stain procedure is useful in determining whether or not the bacteria in question produces tough, dormant spores.

Differential & Selective Media

Media is an agar-based substance used to grow bacteria in the laboratory environment. Some types of agar will grow just about anything, whereas others are differential or selective media.

Selective media excludes some types of bacteria, whereas differential media allows for a colour change that reveals bacteria that have specific metabolic capabilities. MacConkey's Agar (MAC) grows only Gram-negative bacteria and contains a differential pH indicator that changes the colour of bacteria that can ferment the sugar lactose. Mannitol Salt (MSA) is a medium with high sodium chloride (NaCl) content, selective for halophile (salt-loving) bacteria. Mannitol salt also has a pH indicator that changes colour when there are bacteria present that can ferment mannitol, a sugar alcohol also known as a polyol. Pathogenic *Staphylococcus aureus* grows well on this medium and produces the colour change of a mannitol fermenter.

Blood Agar (BAP) is nutrient-rich and will grow many different types of bacteria, but the presence of sheep's blood in the agar allows for a differential distinction between organisms possessing hemolytic enzymes that can cut up red blood cells. This medium is used to test for beta-hemolytic bacteria, like the pathogenic *Streptococcus pyogenes. S. pyogenes* completely destroys the red blood cells in BAP, resulting in clear halos around its bacterial colonies and, if obtained from a throat culture, a diagnosis of strep throat.

Staphylococcus

The *Staphylococcus* genus include just thirty-three species. Most are harmless and reside normally on the skin and mucous membranes of humans and other organisms. Found worldwide, they are a small component of soil microbial flora.

Role in Disease

Staphylococcus can cause a wide variety of diseases in humans and other animals through either toxin production or penetration. Staphylococcal toxins are a common cause of food poisoning, as it can grow in improperly-stored food.

Classification

The main classification of staphylococci is based on their ability to produce coagulase, an enzyme that causes blood clot formation.

Coagulase-positive

- *S. aureus* is coagulase-positive, meaning that they can produce coagulase. However, while the majority of S. aureus are coagulase-positive, some may be atypical in that they do not produce coagulase. S aureus is also catalase-positive (meaning that it can produce the enzyme catalase) and able to convert hydrogen peroxide (H_2O_2) to water and oxygen, which makes the catalase test useful to distinguish Staphylococci from Enterococci and Streptococci.

- *S. pseudintermedius* inhabits and sometimes infects the skin of domestic dogs and cats. This organism, too, can carry the genetic material that imparts multiple bacterial resistance. It is rarely implicated in infections in humans, as a zoonosis.

Coagulase-negative

- *S. epidermidis*, a coagulase-negative staphylococcus species, is a commensal of the skin, but can cause severe infections in immune-suppressed patients and those with central venous catheters.
- *S. saprophyticus*, another coagulase-negative species that is part of the normal vaginal flora, is predominantly implicated in genitourinary tract infections in sexually-active young women.
- In recent years, several other *Staphylococcus* species have been implicated in human infections, notably *S. lugdunensis*, *S. schleiferi*, and *S. caprae*.

Common abbreviations for coagulase-negative staphylococcus species are CoNS and CNS.

Genomics and Molecular Biology

The first S. aureus genomes to be sequenced were those of N315 and Mu50 in 2001. Many more complete S. aureus genomes have been submitted to the public databases, making S. aureus one of the most extensively sequenced bacteria. The use of genomic data is now widespread and provides a valuable resource for researchers working with S. aureus. Whole genome technologies such as sequencing projects and microarrays have shown there is an enormous variety of S. aureus strains. Each contains different combinations of surface proteins and different toxins.

Relating this information to pathogenic behaviour is one of the major areas of staphylococcal research. The development of molecular typing methods has enabled the tracking of different strains of S. aureus. This may lead to better control of outbreak strains. A greater understanding of how the staphylococci evolve, especially due to the acquisition of mobile genetic elements encoding resistance and virulence genes is helping to identify new outbreak strains and may even prevent their emergence.

Staphylococcal Infection

Staphylococcal toxins are a common cause of food poisoning, as it can grow in improperly-stored food.

Coagulase-positive

These bacteria can survive on dry surfaces, increasing the chance of transmission. S. aureus is also implicated in toxic shock syndrome; during the 1980s some tampons allowed the rapid growth of S. aureus, which released toxins that were absorbed into the bloodstream. Any S. aureus infection can cause the staphylococcal scalded skin syndrome, a cutaneous reaction to exotoxin absorbed into the bloodstream. It can also cause a type of septicaemia called pyaemia.

The infection can be life-threatening. Problematically, Methicillin-resistant Staphylococcus aureus (MRSA) has become a major cause of hospital-acquired infections, and is being recognized with increasing frequency in community-acquired infections.

Table: *Main Staphylococcus Aureus Infections*

Type	***Examples***
Localized skin infections	• Stye and other small, superficial abscesses in sweat or sebaceous glands · Subcutaneous abscesses (boils) around foreign bodies
	• Large, deep, infections (carbuncles) possibly causing bacteremia
Diffuse skin infection	• Impetigo
Deep, localized infections	Acute and chronic osteomyelitis
	• Septic arthritis
Other infections	• Acute infective endocarditis
	• Septicemia
	• Necrotizing pneumonia
Toxinoses	• Toxic shock syndrome
	• Gastroenteritis
	• Scalded skin syndrome

Staphylococcus Aureus

Staphylococcus aureus is the most common cause of staph infections. It is a spherical bacterium, frequently part of the skin flora found in the nose and on skin. About 20% of the population are long-term carriers of S. aureus. S. aureus can cause a range of illnesses from minor skin infections, such as pimples, impetigo (may also be caused by Streptococcus pyogenes), boils (furuncles), cellulitis folliculitis, carbuncles, scalded skin syndrome and abscesses, to life-threatening diseases such as pneumonia, meningitis, osteomyelitis, endocarditis, toxic shock syndrome (TSS), bacteremia and sepsis.

Its incidence is from skin, soft tissue, respiratory, bone, joint, endovascular to wound infections. It is still one of the five most common causes of nosocomial infections, often causing postsurgical wound infections. Abbreviated to S. aureus or Staph aureus in medical literature, S. aureus should not be confused with the similarly named and similarly dangerous (and also medically relevant) species of the genus Streptococcus. S. aureus was discovered in Aberdeen, Scotland in 1880 by the surgeon Sir Alexander Ogston in pus from surgical abscesses. Each year some 500,000 patients in American hospitals contract a staphylococcal infection.

Microbiology

S. aureus is a facultatively anaerobic, Gram-positive coccus, which appears as grape-like clusters when viewed through a microscope and has large, round, golden-yellow colonies, often with hemolysis, when grown on blood agar plates. The golden appearance is the etymological root of the bacteria's name; aureus means "golden" in Latin. S. aureus is catalase positive (meaning that it can produce the enzyme "catalase") and able to convert hydrogen peroxide (H_2O_2) to water and oxygen, which makes the catalase test useful to distinguish staphylococci from enterococci and streptococci. A small percentage of S. aureus can be differentiated from most other staphylococci by the coagulase test: S. aureus is primarily coagulase-positive (meaning that it can produce "coagulase", a protein product, which is an enzyme) that causes clot formation while most other Staphylococcus species are coagulase-negative. However, while the majority of S. aureus are coagulase-positive, some may be atypical in that they do not produce coagulase (the most common organism in patients with nosocomial bacteremia is coagulase-negative staphylococcus). Incorrect identification of an isolate can impact implementation of effective treatment and/or control measures.

Role in Disease

Strains are responsible for food poisoning through the production of an enterotoxin and pathogenicity is also associated with coagulase positivity. S. aureus may occur as a commensal on human skin; it also occurs in the nose frequently (in about a third of the population) and throat less commonly. The occurrence of S. aureus under these circumstances does not always indicate infection and therefore does not always require treatment (indeed, treatment may be ineffective and re-colonisation may occur). It can survive on domesticated animals such as dogs, cats and horses, and can cause bumblefoot in chickens. It can survive for some hours on dry environmental surfaces, but the importance of the environment in spread of S. aureus is currently debated. It can host phages, such as the Panton-Valentine leukocidin, that increase its virulence.

S. aureus can infect other tissues when barriers have been breached (e.g., skin or mucosal lining). This leads to furuncles (boils) and carbuncles (a collection of furuncles). In infants S. aureus infection can cause a severe disease Staphylococcal scalded skin syndrome (SSSS). S. aureus alv infections can be spread through contact with pus from an infected wound, skin-to-skin contact with an infected person by producing hyaluronidase that destroy tissues, and contact with objects such as towels, sheets, clothing, or athletic equipment used by an infected person. Deeply penetrating S. aureus infections can be severe. Prosthetic joints put a person at particular risk for septic arthritis, and staphylococcal endocarditis (infection of the heart valves) and pneumonia, which may be rapidly spread.

Atopic Dermatitis

S. aureus is extremely prevalent in atopic dermatitis patients, who are less resistant to it than other people. It often causes complications. The disease is most likely found in fertile active places including, the armpits, hair and scalp. The large pimples that appear in those areas may cause the worst of the infection if popped. This can lead to Scalded skin syndrome. A severe form of this is Ritter's disease seen in neonates.

Toxic Shock Syndrome and S. Aureus Food Poisoning

Some strains of *S. aureus*, which produce the exotoxin TSST-1, are the causative agents of toxic shock syndrome. Some strains of *S. aureus* also produce an enterotoxin that is the causative agent of *S. aureus* gastroenteritis. The gastroenteritis is self-limiting with the person getting better in 8–24 hours. Symptoms include nausea, vomiting, diarrhea, and abdominal pain.

Virulence Factors

Toxins

Depending on the strain, *S. aureus* is capable of secreting several toxins, which can be categorized into three groups. Many of these toxins are associated with specific diseases.

Pyrogenic Toxin Superantigens

(PTSAgs) have superantigen activities that induce toxic shock syndrome (TSS). This group includes the toxin TSST-1, which causes TSS associated with tampon use. The staphylococcal enterotoxins, which cause a form of food poisoning, are included in this group.

Exfoliative Toxins

EF toxins are implicated in the disease staphylococcal scalded-skin syndrome (SSSS), which occurs most commonly in infants and young children. It also may occur as epidemics in hospital nurseries. The protease activity of the exfoliative toxins causes peeling of the skin observed with SSSS.

Other Toxins

Staphylococcal toxins that act on cell membranes include alpha-toxin, beta-toxin, delta-toxin, and several bicomponent toxins. The bicomponent toxin Panton-Valentine leukocidin (PVL) is associated with severe necrotizing pneumonia in children. The genes encoding the components of PVL are encoded on a bacteriophage found in community-associated MRSA strains.

Protein A

Protein A is a protein that is anchored to staphylococcal peptidoglycan pentaglycine bridges by the transpeptidase Sortase A. Protein A is an IgG-binding protein which binds to the Fc region of an antibody. In fact, studies involving mutation of genes coding for Protein A resulted in a lowered virulence of *S. aureus* as measured by survival in blood, and this has led to speculation that Protein A contributed virulence requires binding of antibody Fc regions. Protein A in various recombinant forms has been used for decades to bind and purify a wide range of antibodies by immunoaffinity chromatography. Transpeptidases such as the sortases which are responsible for anchoring factors like Protein A to the staphylococcal peptidoglycan are being studied in hopes of developing new antibiotics to target MRSA infections.

Role of Pigment in Virulence

Some strains of *S. aureus* are capable of producing *staphyloxanthin*- a carotenoid pigment that acts as a virulence factor. Its has an antioxidant action that helps the microbe to evade killing with reactive oxygen used by the host immune system. It is thought that staphyloxanthin is responsible for S. aureus' characteristic golden colour. When comparing a normal strain of *S. aureus* with a strain modified to lack the yellow coloration, the pigmented strain was more likely to survive dousing with an oxidizing chemical such as hydrogen peroxide than the mutant strain was.

Colonies of the two strains were also exposed to human neutrophils. The mutant colonies quickly succumbed while many of the pigmented colonies survived. Wounds on mice were swiped with the two strains. The pigmented strains created lingering abscesses. Wounds with the unpigmented strains healed quickly. These tests suggest that the yellow pigment may be key to the ability of *S. aureus* to survive immune system attacks. Drugs designed to inhibit the bacterium's production of the staphyloxanthin may weaken it and renew its susceptibility to antibiotics. In fact, because of similarities in the pathways for biosynthesis of staphyloxanthin and human cholesterol, a drug developed in the context of cholesterol-lowering therapy was shown to block S. aureus pigmentation and disease progression in a mouse infection model.

Diagnosis

Depending upon the type of infection present, an appropriate specimen is obtained accordingly and sent to the laboratory for definitive identification by using biochemical or enzyme-based tests. A Gram stain is first performed to guide the way, which should show typical gram-positive bacteria, cocci, in clusters. Secondly, culture the organism in mannitol salt agar, which is a selective medium with 7–9% NaCl that allows S. aureus to grow producing yellow-colored colonies as a result of mannitol fermentation and subsequent drop in the medium's pH. Furthermore, for differentiation on the species level, catalase (positive for all Staphylococcus species), coagulase (fibrin clot formation, positive for S. aureus), DNAse (zone of clearance on nutrient agar), lipase (a yellow colour and rancid odour smell), and phosphatase (a pink colour) tests are all done. For staphylococcal food poisoning, phage typing can be performed to determine if the staphylococci recovered from the food to determine the source of infection.

Rapid Diagnosis and Typing

Diagnostic microbiology laboratories and reference laboratories are key for identifying outbreaks and new strains of *S. aureus*. Recent genetic advances have enabled reliable and rapid techniques for the identification and characterization of clinical isolates of *S. aureus* in real-time. These tools support infection control strategies to limit bacterial spread and ensure the appropriate use of antibiotics. These techniques include Real-time PCR and Quantitative PCR and are increasingly being employed in clinical laboratories.

Treatment and Antibiotic Resistance

The treatment of choice for S. aureus infection is penicillin; but in most countries, penicillin-resistance is extremely common and first-line therapy is most commonly a penicillinase-resistant penicillin (for example, oxacillin or flucloxacillin). Combination therapy with gentamicin may be used to treat serious infections like endocarditis, but its use is controversial because of the high risk of damage to the kidneys. The duration of treatment depends on the site of infection and on severity.

Antibiotic alvz resistance in S. aureus was almost unknown when penicillin was first introduced in 1943; indeed, the original petri dish on which Alexander Fleming of Imperial College London observed the antibacterial activity of the penicillium mould was growing a culture of S. aureus. By 1950, 40% of hospital S. aureus isolates were penicillin resistant; and by 1960, this had risen to 80%. Researchers from Italy have identified a bacteriophage active against Staphylococcus aureus, including methicillin-resistant strains, in mice and possibly humans.

Mechanisms of Antibiotic Resistance

Staphylococcal resistance to penicillin is mediated by penicillinase (a form of â-lactamase) production: an enzyme which breaks down the â-lactam ring of the penicillin molecule. Penicillinase-resistant penicillins such as methicillin, nafcillin, oxacillin, cloxacillin, dicloxacillin and flucloxacillin are able to resist degradation by staphylococcal penicillinase. Resistance to methicillin is mediated via the *mec* operon, part of the staphylococcal cassette chromosome mec (SCC*mec*). Resistance is conferred by the *mecA* gene, which codes for an altered penicillin-binding protein (PBP2a or PBP2') that has a lower affinity for binding â-lactams (penicillins, cephalosporins and carbapenems). This allows for resistance to all â-lactam antibiotics and obviates their clinical use during MRSA infections. As such the glycopeptide, vancomycin, is often deployed against MRSA. Aminoglycosides such as kanamycin, gentamicin, streptomycin, etc. were once effective against Staphylococcal infections until the organism evolved mechanisms to destroy the aminoglycosides action, which occurs via protonated amine and/or hydroxyl interactions with the ribosomal RNA of the bacterial 30S Ribosome There are three main mechanisms of aminoglycoside resistance mechanisms which are currently and widely accepted: Aminoglycoside modifying enzymes, Ribosomal mutations, and active efflux of the drug out of the bacteria.

Aminoglycoside modifying enzymes are enzymes that inactivate the aminoglycoside by covalently attaching either a phosphate, nucleotide, or acetyl moiety to either the amine and/or alcohol functionality of the antibiotic; thus rendering the antibiotic through sterics or lack of charge, ineffective in ribosomal binding affinity. In *Staphylococcus Aureus* the best characterized aminoglycoside modifying enzyme is ANT(4')IA *Aminoglycoside adenylyltransferase 4' IA*. This enzyme has been solved by X-Ray Crystallography The enzyme is able to attach an adenyl moiety to the 4' hydroxyl group of many aminoglycosides including kamamycin and gentamicin.

Glycopeptide resistance is mediated by acquisition of the vanA gene. The vanA gene originates from the enterococci and codes for an enzyme that produces an alternative peptidoglycan to which vancomycin will not bind. Today, *S. aureus* has become resistant to many commonly used antibiotics. In the UK, only 2% of all *S. aureus* isolates are sensitive to penicillin with a similar picture in the rest of the world, due to a penicillinase (a form of â-lactamase). The â-lactamase-resistant penicillins (methicillin, oxacillin, cloxacillin and flucloxacillin) were developed to treat penicillin-resistant *S. aureus* and are still used as first-line treatment. Methicillin was the first antibiotic in this class to be used (it was introduced in 1959), but only two years later, the first case of methicillin-resistant *S. aureus* (MRSA) was reported in England.

Despite this, MRSA generally remained an uncommon finding even in hospital settings until the 1990s when there was an explosion in MRSA prevalence in hospitals where it is now endemic. MRSA infections in both the hospital and community setting are commonly treated with non-â-lactam antibiotics such as clindamycin (a lincosamine) and co-trimoxazole (also commonly known as trimethoprim/sulfamethoxazole). Resistance to these antibiotics has also led to the use of new, broad-spectrum anti-Gram positive antibiotics such as linezolid because of its availability as an oral drug. First-line treatment for serious invasive infections due to MRSA is currently glycopeptide antibiotics (vancomycin and teicoplanin). There are number of problems with these antibiotics, mainly centred around the need for intravenous administration (there is no oral preparation available), toxicity and the need to monitor drug levels regularly by means of blood tests. There are also concerns that glycopeptide antibiotics do not penetrate very well into infected tissues (this is a particular concern with infections of the brain and meninges and in endocarditis). Glycopeptides must not be used to treat methicillin-sensitive *S. aureus* (MSSA) as outcomes are inferior.

Because of the high level of resistance to penicillins, and because of the potential for MRSA to develop resistance to vancomycin, the Centers for Disease Control and Prevention have published guidelinesfor the appropriate use of vancomycin. In situations where the incidence of MRSA infections is known to be high, the attending physician may choose to use a glycopeptide antibiotic until the identity of the infecting organism is known. When the infection is confirmed to be due to a methicillin-susceptible strain of *S. aureus*, then treatment can be changed to flucloxacillin or even penicillin as appropriate.

Vancomycin-resistant *S. aureus* (VRSA) is a strain of *S. aureus* that has become resistant to the glycopeptides. The first case of vancomycin-intermediate *S. aureus* (VISA) was reported in Japan in 1996; but the first case of *S. aureus* truly resistant to glycopeptide antibiotics was only reported in 2002. Three cases of VRSA infection have been reported in the United States as of 2005.

Infection Control

' Spread of S. aureus (including MRSA) is through human-to-human contact, although recently some vets have discovered that the infection can be spread through pets, with environmental contamination thought to play a relatively unimportant part. Emphasis on basic hand washing techniques are therefore effective in preventing the transmission of S. aureus. The use of disposable aprons and gloves by staff reduces skin-to-skin contact and therefore further reduces the risk of transmission. Please refer to the article on infection control for further details.

Recently, there have been myriad reported cases of S. aureus in hospitals across America. The pathogen has had facilitated transportation in medical facilities mainly because of insufficient healthcare worker hygiene. S. aureus is an incredibly hardy bacterium, as was shown in a study where it survived on a piece of polyester for just under three months, polyester being the main material used in hospital privacy curtains. The bacterium is able to transport itself on the hands of healthcare workers who, for instance, get the bacteria from a seemingly healthy patient carrying a "benign" or commensal strain of the pathogen and then pass it on to the next patient being cared for. Introduction of the bacterium into the bloodstream can lead to various complications including, but not limited to, endocarditis, meningitis, and, if it is widespread, sepsis- toxins infecting the entire body.

Because of these infections in hospitals, as of February 14, 2008, all California medical facilities must now report S. aureus infections that are checked into the hospitals, in the hope of starting a trend to aid disease trackers and pathologists in their search for a cure. Alcohol has proven to be an effective topical sanitizer against MRSA. Quaternary ammonium can be used in conjunction with alcohol to increase the duration of the sanitizing action. The prevention of nosocomial infections involve routine and terminal cleaning. Nonflammable alcohol vapor in CO_2 NAV-CO2 systems have an advantage as they do not attack metals or plastics used in medical environments, and do not contribute to antibacterial resistance.

An important and previously unrecognized means of community-associated methicillin-resistant S. aureus colonization and transmission is during sexual contact. Staff or patients who are found to carry resistant strains of S. aureus may be required to undergo "eradication therapy" which may include antiseptic washes and shampoos (such as chlorhexidine) and application of topical antibiotic ointments (such as mupirocin or neomycin) to the anterior nares of the nose. The nonprotein amino acid L-homoarginine is a growth inhibitor of Staphylococcus aureus as well as Candida albicans. It is assumed to be an antimetabolite of arginine. In March 2007, the BBC reported that a vaporizer spraying some essential oils (including tea tree oil) into the atmosphere reduced airborne bacterial counts by 90% and kept MRSA infections at bay and may hold promise in MRSA infection control. '"

Staphylococcus Aureus and Staphylococcal Disease

Staphylococci (staph) are Gram-positive spherical bacteria that occur in microscopic clusters resembling grapes. Bacteriological culture of the nose and skin of normal humans invariably yields staphylococci.

In 1884, Rosenbach described the two pigmented colony types of staphylococci and proposed the appropriate nomenclature: *Staphylococcus aureus* (yellow) and *Staphylococcus albus* (white). The latter species is now named *Staphylococcus epidermidis.* Although more than 20 species of *Staphylococcus* are described in Bergey's Manual (2001), only *Staphylococcus aureus* and *Staphylococcus epidermidis* are significant in their interactions with humans. *S. aureus* colonizes mainly the nasal passages, but it may be found regularly in most other anatomical locales, including the skin, oral cavity and gastrointestinal tract. *S epidermidis* is an inhabitant of the skin.

Taxonomically, the genus *Staphylococcus* is in the Bacterial family *Staphylococcaceae,* which includes three lesser known genera, *Gamella, Macrococcus* and *Salinicoccus.* The best-known of its nearby phylogenetic relatives are the members of the genus *Bacillus* in the family *Bacillaceae,* which is on the same level as the family *Staphylococcaceae.* The *Listeriaceae* are also a nearby family.

Staphylococcus aureus forms a fairly large yellow colony on rich medium; *S. epidermidis* has a relatively small white colony. *S. aureus* is often hemolytic on blood agar; *S. epidermidis* is non hemolytic. Staphylococci are facultative anaerobes that grow by aerobic respiration or by fermentation that yields principally lactic acid. The bacteria are catalase-positive and oxidase-negative. *S. aureus* can grow at a temperature range of 15 to 45 degrees and at NaCl concentrations as high as 15 percent. Nearly all strains of *S. aureus* produce the enzyme coagulase: nearly all strains of *S. epidermidis* lack this enzyme. *S. aureus* should always be considered a potential pathogen; most strains of *S. epidermidis* are nonpathogenic and may even play a protective role in humans as normal flora. *Staphylococcus epidermidis* may be a pathogen in the hospital environment.

Staphylococci are perfectly spherical cells about 1 micrometer in diameter. The staphylococci grow in clusters because the cells divide successively in three perpendicular planes with the sister cells remaining attached to one another following each successive division. Since the exact point of attachment of sister cells may not be within the divisional plane, and the cells may change position slightly while remaining attached, the result is formation of an irregular cluster of cells. The shape and configuration of the Gram-positive cocci helps to distinguish staphylococci from streptococci. Streptococci are slightly oblong cells that usually grow in chains because they divide in one plane only, similar to a bacillus. Without a microscope, the catalase test is important in distinguishing streptococci (catalase-negative) from staphylococci, which are vigorous catalase-producers. The test is performed by adding 3% hydrogen peroxide to a colony on an agar plate or slant. Catalase-positive cultures produce O_2 and bubble at once. The test should not be done on blood agar because blood itself contains catalase. Pathogenesis of *S. aureus* infections

Staphylococcus aureus causes a variety of suppurative (pus-forming) infections and toxinoses in humans. It causes superficial skin lesions such as boils, styes and furuncules; more serious infections such as pneumonia, mastitis, phlebitis, meningitis, and urinary tract infections; and deep-seated infections, such as osteomyelitis and endocarditis. *S. aureus* is a major cause of hospital acquired (nosocomial) infection of surgical wounds and infections associated with indwelling medical devices. *S. aureus* causes food poisoning by releasing enterotoxins into food, and toxic shock syndrome by release of superantigens into the blood stream. Although methicillin-resistant Staph aureus (MRSA) have been entrenched in hospital settings for several decades, MRSA strains have recently emerged outside the hospital becoming known as community associated- MRSA((CA-MRSA) or superbug strains of the organism, which now account for the majority of staphylococcal infections seen in the ER or clinic.

S. aureus expresses many potential virulence factors: (1) surface proteins that promote colonization of host tissues; (2) invasins that promote bacterial spread in tissues (leukocidin, kinases, hyaluronidase); (3) surface factors that inhibit phagocytic engulfment (capsule, Protein A); (4) biochemical properties that enhance their survival in phagocytes (carotenoids, catalase production); (5) immunological disguises (Protein A, coagulase, clotting factor); (6) membrane-damaging toxins that lyse eucaryotic cell membranes (hemolysins, leukotoxin, leukocidin; (7) exotoxins that damage host tissues or otherwise provoke symptoms of disease (SEA-G, TSST, ET); and (8) inherent and acquired resistance to antimicrobial agents

For the majority of diseases caused by *S. aureus*, pathogenesis is multifactorial, so it is difficult to determine precisely the role of any given factor. However, there are correlations between strains isolated from particular diseases and expression of particular virulence determinants, which suggests their role in a particular diseases. The application of molecular biology has led to advances in unraveling the pathogenesis of staphylococcal diseases. Genes encoding potential virulence factors have been cloned and sequenced, and many protein toxins have been purified. With some staphylococcal toxins, symptoms of human disease can be reproduced in animals with the purified protein toxins, lending an understanding of their mechanism of action.

Human staphylococcal infections are frequent, but usually remain localized at the portal of entry by the normal host defences. The portal may be a hair follicle, but usually it is a break in the skin which may be a minute needle-stick or a surgical wound. Foreign bodies, including sutures, are readily colonized by staphylococci, which may make infections difficult to control. Another portal of entry is the respiratory tract. Staphylococcal pneumonia is a frequent complication of influenza.

The localized host response to staphylococcal infection is inflammation, characterized by an elevated temperature at the site, swelling, the accumulation of pus, and necrosis of tissue. Around the inflamed area, a fibrin clot may form, walling off the bacteria and leukocytes as a characteristic pus-filled boil or abscess. More serious infections of the skin may occur, such as furuncles or impetigo. Localized infection of the bone is called osteomyelitis. Serious consequences of staphylococcal infections occur when the bacteria invade the blood stream. A resulting septicemia may be rapidly fatal; a bacteremia may result in seeding other internal abscesses, other skin lesions, or infections in the lung, kidney, heart, skeletal muscle or meninges.

Staphylococcus

Higher Order Taxa

Bacter7ia; Firmicutes; Bacilli; Bacillales; Staphylococcaceae; Staphylococcus.

Species: Staphylococcus 9IX/A01/138, S. arlettae, S. aureus, S. auricularis, S. capitis, S. caprae, S. carnosus, S. chromogenes, S. cohnii, S. condimenti, S. croceolyticus, S. delphini, S. epidermidis, S. equorum, S. faecalis, S. felis, S. fleurettii, S. gallinarum, *S. haemolyticus*, S. hominis, S. hyicus, S. intermedius, S. kloosii, S. lentus, S. lugdunensis, S. lutrae, S. lyticans, S. muscae, S. nepalensis, S. pasteuri, S. pettenkoferi, S. piscifermentans, S. pseudintermedius, S. roseus, S. saccharolyticus, S. saprophyticus, S. schleiferi, S. sciuri, S. simiae, S. simulans, S. succinus, S. vitulinus, S. warneri, S. xylosus.

Staphylococci are spherical Gram-positive bacteria, which are immobile and form grape-like clusters. They form bunches because they divide in two planes as opposed to their close relatives streptococci which, although they are similarly shaped, form chains because they divide only in one plane. Staphylococci are facultative anaerobes. They mainly grow by aerobic respiration, or fermentation that produces lactic acid. There are two different types of colonies formed. Those formed by *S. aureus* are yellow and become rather large on a rich medium, while those of *S. epidermidi* are white and form relatively small colonies, even on a rich medium. As a pathogen, it is important to understand the virulence mechanisms of *Staphylococcus* in order to successfully combat the pathogen. Strains of *Staphylococcus* resistant to our most powerful antibiotics have increased pressure on researchers to find alternative, more effective ways of fighting *Staphylococcus*.

Genome Structure

There are currently 11 genome projects, 9 of which are complete and 2 are in process. The Staphylococcus aureus genome, which is the most common species amoung the genome projects, is about 2.8 Mega-base pairs long with approximately 2,600 open reading frames, comprising 84.5% of the genome. It has a G+C content of about 33%, and a single plasmid about 25,000 base pairs in length with a G+C content of 28.7%. The genes for antibiotic resistance in Staphylococcus aureus are located on plasmids or other similar structures. Staphylococcus epidermidis, which has two genomes sequenced in the speices, is about 2.5 Mega-base pairs long with 1,681 open reading frames and its G+C content is 32%. Staphylcoccus haemolyticus is about 2.68 Mega-base pairs long with a G+C content of 32%.

Virulence factors are encoded by phages, plasmids, pathogenicy islands and staphylococcus cassette chromosome. Increased resistance for antibiotics is encoded by a transposon (Tn 1546) that was inserted into a conjugated plasmid that also encoded resistance to other things including disinfectants. MRSA (Methicillin-Resistant Staphylococcus Aureus), which is resistant to the antibiotic methicillin, expresses a modified penicillin-binding protein encoded by mecA gene. This was brought about by many evolutions thought horizontal gene transfer of mecA to a wide variety of methicillin susceptible S. Aureus strains. Diversification within the S. Aureus population is achieved through a combination of mutation, recombination and horizontal gene transfer. Evolution of this bacteria can occur through asymptomatic colonization and/or during the course of the caused disease.

Cell Structure and Metabolism

Staphylococcus Aureus is a gram positive bacteria, which means that the cell wall of this bacteria consists of a very thick peptidoglycan layer. They are spherical, form clusters in 2 planes and have no flagella. Secretions are numerous, but include surface associated adhesins, endotoxins, exoenzymes, capsular polysaccharide. the capsule is responsible for enhanced virulence of a mucoid strain.

Ecology

Staphylococcus Aureus is a normal inhabitant of the skin and mucous membranes in the nose of a healthy human, while S. epidermidis inhabits only the skin of healthy humans. S. Aureus is infectious to both animals and humans. Approximately 30% of the normal healthy population is affected by S. Aureus as it asymptomatically colonizes on human hosts.

Pathology

Staphylococcus Aureus is the most common cause of staph infections and is responsible for various diseases including: mild skin infections (impetigo, folliculitis, etc.), invasive diseases (wound infections, osteomyelitis, bacteremia with metastatic complications, etc.), and toxin mediated diseases (food poisoning, toxic shock syndrome or TSS, scaled skin syndrome, etc.). Infections are preceded by colonization. Common superficial infections include carbuncles, impetigo, cellulitis, folliculitis. Community-acquired infections influde bacteremia, endocarditis, osteomylitis, pneumonia and wound infections are less common. S. Aureus also cuases economically importatnt mastitis in cows, sheep and goats.

Staphylococcus usually only infects immunocompromised humans, and is an extremely versatile pathogen. Infections usually remain localized at the point of entry by host defences. S. aureus is one of the major causes of hospital-acquired infection. Foreign bodies, such as sutures, indwelling catheters, and implanted joints, are extremely susceptible to Staphylococcus epidermidis colonization, and often serve as the point of entry for infection. Another common path of infection is the respiratory tract, where the bacterium causes pneumonia. By forming biofilms Staphylococcus epidermidis is resistant to antibiotics, antibodies, and phagocytosis, and can serve as a reservoir for antibiotic resistant genes that can be transferred to other bacteria. Other diseases caused by staphylococci include boils, sinusitis, emesis, diarrhea, endocarditis, scalded skin syndrome, osteomyelitis, urinary tract infection, and toxic shock syndrome. In the late 1970's an epidemic of toxic shock syndrome (TSS T-1) was brought about by a change in host environment. The environment change was encouraged by a creation of modern technology otherwise known as the super-absorbent tampon. This new modern convenience product that had swept the nation's females actually created a new nutrient-rich surface area for many bacteria to thrive off of, "S.

Aureus" being a major resident. However, TSS from tampons can be easily avoided by correctly using tampons (read instructions and warning labels listed with tampon products).

The first serious emergence of antibiotic resistance staph occurred with a specific strain we refer to as Methicillin-Resistant Staphylococcus Aureus, abbreviated as MRSA. This strain expressed a modified penicillin-binding protein encoded by mecA gene and is present in 4 forms of staph. cassette. The MRSA, resistant to the antibiotic methicillin, and eventually was isolated.

Consequently, vancomycin (the most powerful antibiotic in our arsenal) became the primary antibiotic used to combat staphylococcus infection. In 1997 a strain of S. aureus resistant to vancomycin was isolated, and people are once again exposed to the threat of untreatable staphylococcus infection.

MRSA strains are currently a very significant healthcare problem. The sequencing of the S. aureus genome will hopefully provide insight into how the organism generates such a variety of toxins, and aid researchers in developing ways of combating the versatile bacterium.

6

Nutrition and Cultivation of Bacteria

In the late 20th Century, numerous authors contributed to an essay on bacterial nutrition and cultivation which has been a part of the continuously-evolving microbiology laboratory manuals utilized at UW-Madison.

These two web pages expand on that which is in the manual presently in use for the Microbiology 102 laboratory course, and much of the original authors' wordage remains in sections II-VI. Included are links to related web material.

The survival of microorganisms in the laboratory, as well as in nature, depends on their ability to grow under certain chemical and physical conditions. An understanding of these conditions enables us to characterize isolates and differentiate between different types of bacteria. Such knowledge can also be applied to control the growth of microorganisms in practical situations.

Media used in the laboratory for the cultivation of bacteria must supply all of the necessary nutrients required for cellular growth and maintenance of the organisms.

A wide variety of culture media is employed by the bacteriologist for the isolation, growth and maintenance of pure cultures and also for the identification of bacteria according to their biochemical and physiological properties. A culture medium must supply suitable carbon and energy sources and other nutrients, sometimes including "growth factors". It is important to note that no one medium will support the growth of all microorganisms. Accordingly, the elements required for the maintenance, growth and reproduction of all organisms will be used by different organisms in different ways. When one prepares a medium for the cultivation of microorganisms, one dissolves various organic and/or inorganic compounds sequentially in pure, distilled water. The importance of water cannot be overestimated.

Water is the universal solvent in which all nutrients must be dissolved and all chemical reactions will take place. It can supply some hydrogen and oxygen in certain chemical reactions. Water containing a significant amount of solutes may not be osmotically compatible or "available" for use by microorganisms, so the concept of water availability needs to be considered.

Nutrition and Cultivation of Bacteria

Solid Media

Agar is the major solidifying agent used in bacteriological media. It is an impure polysaccharide gum obtained from certain marine algae. It is added as a powder at a more or less standard concentration (1.5% for plates and slanted media, 0.5% or less for "semisolid" media), usually after the other medium components have been added and dissolved in the water. Agar dissolves at approximately 100°C, and an agar-containing medium thus heated will not solidify until the temperature is brought down to about 43°C.

Once solidified, the medium will not melt until brought back up to about 100°C. Among the advantages of this interesting temperature-related property are the following: (1) The medium can be inoculated while in a liquid state at a low enough temperature (approx. 43-50°C) such that the cells will not die off, and (2) the medium, once solidified, will stay solid over a wide range of incubation conditions.

Two additional attributes of agar are its resistance to degradation by nearly all organisms and its relative clarity, permitting easy viewing of growth on or in the medium.

One drawback to agar is the fact that it is very difficult, if not impossible, to purify it fully of trace impurities. Thus, when agar is added to a chemically-defined liquid medium, the medium must be considered complex. If an absolutely chemically-defined solid medium is required, silicon-based solidifying agents can be employed.

Previous to agar, potato slices and gelatin were utilized to form solid substrates upon which microbial colonies could be grown and studied. These materials were unacceptable for general use due to their ability to be broken down by a wide variety of microorganisms.

Furthermore, gelatin liquefies in a warm room, and potato slices are opaque. In 1881, Fanny Eilshemius Hesse, a technician in the laboratory of Robert Koch in Germany, introduced the concept of agar to bacteriology, having used it for many years in the preparation of homemade jellies.

Classification of Culture Media

A classification of media based on their respective uses follows. Note that these categories can overlap. Furthermore, by now you should be using these terms correctly: Medium is always the singular form of the word, and media is always (and only!) the plural form.

- A Minimal Medium is one which supplies only the minimal nutritional requirements of a particular organism. As an example, a typical, prototrophic strain of *E. coli* is able to synthesize all of its cell components from a simple solution containing several "mineral salts" plus glucose as the source of carbon and energy – such as the medium given on the previous page. Minimal media vary in composition according to the minimal nutritional requirements of the particular species under study. A minimal medium for a "fastidious" organism such as *Lactobacillus* may contain many growth factors such as vitamins and amino acids.
- An all-purpose Medium is rich in a wide variety of nutrients (including many growth factors) and will, therefore, support the growth of a wide range of bacteria. All-purpose media include Nutrient Agar, APT Agar, Plate Count Agar, Heart Infusion Agar, Brain Heart Infusion Agar and Penassay Agar.
- A Selective Medium supports the growth of desired organisms while inhibiting the growth of many or most of the unwanted ones – either by purposely adding one or more selective agents which "poison" certain types of organisms or by including or deleting certain nutrients such that the desired organisms and few others are able to grow. Examples on how these things may be accomplished are as follows:
    - o *MacConkey Agar*. This is an example of a medium where selective agents are added which directly suppress the growth of undesired organisms as much as possible. The particular selective agents chosen for this medium – bile salts and crystal violet – inhibit gram-positive bacteria, allowing the near-exclusive growth of gram-negative bacteria.

- *Nitrogen-Free Broth.* Here the medium is made selective by the deletion of a required element; no nitrogen compounds are present. Therefore, the only organisms which can grow after inoculation into this medium are those which can utilize gaseous nitrogen (N_2) which diffuses in from the atmosphere. These are the nitrogen-fixing bacteria. While this medium does not utilize selective agents, it is still restrictive to an extensive number of various organisms.
- *Succinate Broth.* In this example, a particular nutrient utilized by the desired organism – and few others – is included as the only carbon source. This medium is used for the enrichment of the purple non-sulfur photosynthetic bacteria; most other organisms tend not to metabolize succinate under the anaerobic conditions utilized. This is another example of a restrictive medium which does not contain selective agents.

- A Differential Medium is one which allows two or more different types of organisms to grow, but it contains dyes and/or other components upon which different organisms act in various ways to produce a variety of end products or effects, often detected by variations in colour. These differences are often very apparent among colonies of a mixed culture growing in a petri dish. Pure cultures, growing in separate tubes of the same differential medium, may also be characterized and differentiated from one another according to a particular biochemical characteristic. Examples of differential media include the following:
  - *MacConkey Agar.* This medium is used in plates. Organisms which ferment the lactose in the medium will lower the pH due to the production of acids. The pH indicator (neutral red) will turn red, and the colonies will consequently have a reddish appearance. Other colonies on the same plate which do not contain lactose-fermenting cells should appear whitish. (As this medium also appears in the above category, it is termed a selective-differential medium.)
  - *Glucose Fermentation Broth.* This medium is used in tubes, usually with Durham tubes. As discussed on the previous page, organisms which ferment the sugar (glucose) will cause the pH indicator to change colour upon production of acidic products. Additionally, if insoluble gas (H_2) is

produced during fermentation, a bubble will be seen in the inverted Durham tube.

- o Examples of differential media which do not involve pH-related reactions include Motility Medium (exploiting a morphological characteristic – i.e., production of flagella), Nutrient Gelatin, Starch Agar, Kligler Iron Agar (for H_2S production) and Blood Agar.

Summary of Commonly-Used Constituents in Microbiological Media

- Agar. Agar is used as a solidifying agent in media. It is an impure polysaccharide gum obtained from certain marine algae. Agar dissolves and melts around 100°C and solidifies around 43°C. Generally agar itself is not used as a nutrient by microorganisms.
- Body Fluids. Whole or defibrinated blood, plasma, serum or other body fluids are frequently added to culture media for the isolation and cultivation of many pathogens. Body fluids contribute many growth factors and/or substances which detoxify certain inhibitors.
- Buffers. These compounds are incorporated to maintain the optimum pH range of the organism. Substances like sodium and potassium phosphates and calcium carbonate prevent marked changes in pH which otherwise would result from microbial production of organic acids or bases. Crude organic preparations such as peptones also act as buffers.
- Extracts. Eucaryotic tissues (yeast, beef muscle, liver, brain, heart, etc.) are extracted by boiling and then concentrated to a paste or dried to a powder. These extracts are frequently used as a source of amino acids, vitamins and coenzymes, including many needed as growth factors by fastidious organisms. Trace elements and other minerals and usually some sugar are also present. (The term infusion refers to the aqueous extracts of these materials used for these purposes without being dried or otherwise concentrated, although "infusion" is sometimes used synonymously with "extract.")
- Peptones. These complex mixtures of organic and inorganic compounds are obtained by digestion of protein-containing tissues of animals and plants such as meat scraps, beef muscle, gelatin, milk protein (casein) and soybean meal. These materials

are then dried down to a powder and made commercially available to microbiology laboratories. Peptones primarily contain peptides and single amino acids. Being crude digests of complex materials, they contain a great variety of other organic and inorganic materials, but they may be deficient in certain minerals and vitamins. Three examples of brand names of peptones are Tryptone (or Trypticase; a pancreatic digest of casein), Phytone (or Soytone; a papaic digest of soybean meal) and simply Peptone (a digest of beef muscle). Peptones are used frequently in conjunction with extracts for the cultivation of fastidious organisms, and a simple peptone solution will support the growth of many organisms. Peptone in a concentration of 0.1% is often used as a diluent.

- pH Indicators. An acid-base indicator is often added to differential media to detect changes in hydrogen ion concentration during the growth of an organism such as in Carbohydrate Fermentation Broth, Kligler Iron Agar, Simmons Citrate Agar, MacConkey Agar and Glucose O/F Medium. Brom-cresol purple, brom-thymol blue and phenol red are commonly used; for each of these, an acidic pH turns the indicator a yellow colour. A list of pH indicators is given here.
- Reducing Agents. Certain chemicals may stimulate growth by reducing the oxidation-reduction potential in the environment. Cystine and thioglycollate are reducing agents often used for the cultivation of anaerobes and are found in the Thioglycollate Medium utilized to determine oxygen relationships.
- Selective Agents. Antimicrobial agents such as crystal violet, bile salts, brilliant green, potassium tellurite, sodium azide and antibiotics can be employed in selective media to suppress or inhibit the growth of certain groups of microorganisms while allowing growth of desired organisms. These agents are usually bacteriostatic.

Nutrition and Growth of Bacteria

Culture Media for the Growth of Bacteria

For any bacterium to be propagated for any purpose it is necessary to provide the appropriate biochemical and biophysical environment. The biochemical (nutritional) environment is made available as a culture medium, and depending upon the special needs of particular bacteria (as well as particular investigators) a large variety and types of culture media have been developed with different purposes and uses. Culture media are employed in the isolation and maintenance of pure cultures of bacteria and are also used for identification of bacteria according to their biochemical and physiological properties.

The manner in which bacteria are cultivated, and the purpose of culture media, varies widely. Liquid media are used for growth of pure batch cultures, while solidified media are used widely for the isolation of pure cultures, for estimating viable bacterial populations, and a variety of other purposes. The usual gelling agent for solid or semisolid medium is agar, a hydrocolloid derived from red algae. Agar is used because of its unique physical properties (it melts at 100°C and remains liquid until cooled to 40°C, the temperature at which it gels) and because it cannot be metabolized by most bacteria. Hence as a medium component it is relatively inert; it simply holds (gels) nutrients that are in aquaeous solution.

Types of Culture Media

Culture media may be classified into several categories depending on their composition or use. A chemically-defined (synthetic) medium is one in which the exact chemical composition is known. A complex (undefined) medium is one in which the exact chemical constitution of the medium is not known. Defined media are usually composed of pure biochemicals off the shelf; complex media usually contain complex materials of biological origin such as blood or milk or yeast extract or beef extract, the exact chemical composition of which is obviously undetermined.

A defined medium is a minimal medium if it provides only the exact nutrients (including any growth factors) needed by the organism for growth. The use of defined minimal media requires the investigator to know the exact nutritional requirements of the organisms in question. Chemically-defined media are of value in studying the minimal nutritional requirements of microorganisms, for enrichment cultures, and for a wide variety of physiological studies.

Complex media usually provide the full range of growth factors that may be required by an organism so they may be more handily used to cultivate unknown bacteria or bacteria whose nutritional requirement are complex (i.e., organisms that require a lot of growth factors, known or unknown). Complex media are usually used for cultivation of bacterial pathogens and other fastidious bacteria.

Most pathogenic bacteria of animals, which have adapted themselves to growth in animal tissues, require complex media for their growth. Blood, serum and tissue extracts are frequently added to culture media for the cultivation of pathogens. Even so, for a few fastidious pathogens such as *Treponema pallidum*, the agent of syphilis, and *Mycobacterium leprae*, the cause of leprosy, artificial culture media and conditions have not been established. This fact thwarts the ability to do basic research on these pathogens and the diseases that they cause. Other concepts employed in the construction of culture media are the principles of selection and enrichment. A selective medium is one which has a component(s) added to it which will inhibit or prevent the growth of certain types or species of bacteria and/or promote the growth of desired species. One can also adjust the physical conditions of a culture medium, such as pH and temperature, to render it selective for organisms that are able to grow under these certain conditions.

A culture medium may also be a differential medium if allows the investigator to distinguish between different types of bacteria based on some observable trait in their pattern of growth on the medium. Thus a selective, differential medium for the isolation of *Staphylococcus aureus,* the most common bacterial pathogen of humans, contains a very high concentration of salt (which the staph will tolerate) that inhibits most other bacteria, mannitol as a source of fermentable sugar, and a pH indicator dye. From clinical specimens, only staph will grow. *S. aureus* is differentiated from *S. epidermidis* (a nonpathogenic component of the normal flora) on the basis of its ability to ferment mannitol.

Mannitol-fermenting colonies *(S. aureus) produce* acid which reacts with the indicator dye forming a colored halo around the colonies; mannitol non-fermenters (*S. epidermidis*) use other non-fermentative substrates in the medium for growth and do not form a halo around their colonies. An enrichment medium employs a slightly different twist. An enrichment medium contains some component that permits the growth of specific types or species of bacteria, usually because they alone can utilize the component from their environment.

However, an enrichment medium may have selective features. An enrichment medium for nonsymbiotic nitrogen-fixing bacteria omits a source of added nitrogen to the medium. The medium is inoculated with a potential source of these bacteria (e.g. a soil sample) and incubated in the atmosphere wherein the only source of nitrogen available is N_2. A selective enrichment medium for growth of the extreme halophile (*Halococcus*) contains nearly 25 percent salt [NaCl], which is required by the extreme halophile and which inhibits the growth of all other procaryotes.

Table: *Minimal medium for the growth of* Bacillus megaterium. *An example of a chemically-defined medium for growth of a heterotrophic bacterium.*

Component	***Amount***	***Function of component***
sucrose	10.0 g	C and energy source
K_2HPO_4	2.5 g	pH buffer; P and K source
KH_2PO_4	2.5 g	pH buffer; P and K source
$(NH_4)2HPO_4$	1.0 g	pH buffer; N and P source
$MgSO_4$ $7H_2O$	0.20 g	S and Mg^{++} source
$FeSO_4$ $7H_2O$	0.01 g	Fe^{++} source
$MnSO_4$ $7H_2O$	0.007 g	Mn^{++} Source
water	985 ml	pH 7.0

Table: *Defined medium (also an enrichment medium) for the growth of* Thiobacillus thiooxidans, *a lithoautotrophic bacterium.*

Component	***Amount***	***Function of component***
NH_4Cl	0.52 g	N source
KH_2PO_4	0.28 g	P and K source
$MgSO_4$ $7H_2O$	0.25 g	S and Mg^{++} source
$CaCl_2$ $2H_2O$	0.07 g	Ca^{++} source
Elemental Sulfur	1.56 g	Energy source
CO_2	5%*	C source
water	1000 ml	pH 3.0

* Aerate medium intermittently with air containing 5% CO_2.

Table: *Complex medium for the growth of fastidious bacteria.*

Component	***Amount***	***Function of component***
Beef extract	1.5 g	Source of vitamins and other growth factors
Yeast extract	3.0 g	Source of vitamins and other growth factors
Peptone	6.0 g	Source of amino acids, N, S, and P
Glucose	1.0 g	C and energy source
Agar	15.0 g	Inert solidifying agent
water	1000 ml	pH 6.6

Table: *Selective enrichment medium for growth of extreme halophiles.*

Component	***Amount***	***Function of component***
Casamino acids	7.5 g	Source of amino acids, N, S and P
Yeast extract	10.0 g	Source of growth factors
Trisodium citrate		3.0 g C and energy source
KCl	2.0 g	K^+ source
$MgSO_4$ 7 H_2O	20.0 g	S and Mg^{++} source
$FeCl_2$	0.023 g	Fe^{++} source
NaCl	250 g	Na^+ source for halophiles and inhibitory to nonhalophiles
water	1000 ml	pH 7.4

Microbial Nutrition

Aim: To describe the diversity in nutrition found in microorganisms.

Learning outcomes:

1. To know the basic nutritional requirements of microorganisms for their growth.
2. To understand the basis of nutritional classification of microorganisms.
3. To know the different kinds of media and medium design strategies used in Microbiology

Microorganisms, like all organisms, have a basic requirement for several essential elements for growth. These are necessary for the biosynthesis of the components of cells as well as their function. Carbon has a pivotal role in the structure of all molecules found in living cells, however, nitrogen, hydrogen, oxygen, sulphur and phosphorus are all essential as well. Of less significance, in respect of the amount required, are many trace nutrients which are predominantly metal ions e.g. Zn, Cu, Mn etc.

Nutritional Classification

Microorganisms are classified on the basis of their utilisation of nutrients as an energy source and as a carbon source. The major level of classification is into chemotrophs and phototrophs. The former derive energy from chemicals and the latter depend primarily on radiant energy. These two major groups are further classified on the basis of the carbon sources utilised.

Chemoautotrophs (Chemolithotrophs)- use inorganic substrates as sources of energy and $C0_2$ as the main source of carbon.

Chemoheterotrophs (Chemorganotrophs)- utilise organic substrates for both needs.

Photoautotrophs (Photolithotrophs)- use light as energy source and $C0_2$ as carbon source.

Photoheterotrophs (photoorganotrophs)- use light as energy source and organic carbon sources.

Not all bacteria can be so precisely categorised, for example, some phototrophic bacteria can also grow as chemotrophs. *Rhodospirillum rubrum* exists as a photoheterotroph in anaerobic conditions but when oxygen is available, and in the dark, it grows as a chemoheterotroph. This important metabolic switch is also reflected in changes in internal cell structure. The vesicles which contain the photosynthetic pigments become disorganised when the organism grows as a chemoheterotroph.

All fungi are all chemoheterotrophic microorganisms. Chemoheterotrophs show a wide diversity with respect to the sources of organic nutrients that are essential for their growth. At the two extremes are the saprotrophs (saprophytes), which utilise nonliving organic matter, and the biotrophs (parasites) which may be totally dependent on another living organism for their nutrition and cause disease (pathogens). In between there are several types of association between this group of microorganisms and another organism. These include mutalistic or symbiotic associations whereby both partners gain benefit, organisms that live as commensals and under some circumstances cause disease, and finally obligate biotrophs that can live both a saprotrophic and biotrophic existence.

The Source and Supply and Nutrients

The natural environments colonised by bacteria and fungi provide a wide variety of nutrients which will support growth of these organisms. In some organisms, described as fastidious, a specific nutritional requirement can be met from only one source. These organisms are generally pathogens and cannot be cultivated in the laboratory in the absence of their host or host tissues.

The mixtures of nutrients used in the Microbiology laboratory to cultivate bacteria and fungi are called media. These range from complex, undefined mixtures to simple defined mixtures where the nature and amount of a nutrient is known.

The former type of medium is used normally for the routine and regular cultivation of an organism, or when cultures are required quickly for some particular use e.g. for genetic experiments. Conversely for biochemical and physiological experiments a defined medium is favoured e.g. where the regulation of synthesis of an enzyme is being investigated.

Media for the Cultivation of Bacteria

The nutritional diversity of bacteria requires the use of a range of different culture media. Thus autotrophs (chemo and photo) require a medium lacking organic carbon. In the case chemoheterotrophs one nutrient is used as the energy source.

A good example of such organisms are the nitrifying bacteria which derive energy from the oxidation of NH_4 ion (*Nitrosomonas*) and $N0_2$ ion (*Nitrobacter*). A defined medium for a chemoheterotroph has a very similar composition, the notable addition being the organic carbon source.

Media for the Cultivation of Fungi

Fungi are also cultivated on defined or complex media. Note the inclusion of organic carbon in both cases, reflecting the chemoheterotrophic nature of fungi, however, in fungal media extracts of plant materials are used in complex media.

Comparison of bacterial and fungal media shows an important difference in the ratio of carbon to nitrogen; bacterial media have a low C:N ratio and fungi a high C:N.

Special Media

Different types of organisms require the use of particular media. For example *anaerobic bacteria* grow only on a medium lacking oxygen. Reducing agents, such as sodium thioglycollate, are commonly used, but most effective is a piece of apparatus called an anaerobe jar.

This is a vessel in which the atmosphere can be made totally oxygen free.

Selective media are used to enhance growth of particular types of microorganisms. These are particularly useful for the isolation of particular types of organism(s) from a very mixed population e.g. the use of a medium containing brilliant green (a dye) which inhibits gram +ve bacteria, for the isolation of gram-ve types.

Differential media are used to distinguish particular types of organisms in a mixed population. MacConkey medium, which contains bile salts and crystal violet dye, is used to differentiate

Enterobacteriaceae (*Salmonella*, *Shigella* and *Escherichia*) from other gram-ve bacteria. An enrichment medium is used to selectively isolate organisms with a particular nutritional requirement of function. For example, if organisms that can hydrolyse chitin are required samples of natural material e.g. soil are spread on a medium containing chitin. Only chitin degrading organisms will grow.

Agar

All media can be used in liquid or solidified form. The latter include agar as an ingredient. This is a complex polysaccharide extracted from marine algae.

It is normally added at a concentration of 1.5%. Agar has important properties being liquid at 100°C but solidifying at 40°C. This allows the incorporation of bacteria into the medium as an alternative to spreading on the medium surface.

7

Fungal and Bacterial Leaf Diseases

Symptoms: Leaf spot, tar spot, leaf blotch, leaf blister and shot hole are symptoms of leaf diseases. Shot hole occurs when dead leaf tissue falls out, leaving a hole.

Cause: A number of leaf diseases that occasionally damage ornamental plants are caused by fungi and bacteria. Many species of fungi cause leaf diseases. Leaf diseases cause leaves to fall prematurely, leaving trees and shrubs without the green factories that produce food for them. Repeated defoliation of trees or shrubs makes them progressively weaker.

Most deciduous plants can withstand several defoliations without serious damage, but broad-leaved and narrow-leaved evergreens usually do not recover from a defoliation. Most fungi and bacteria that cause leaf diseases require a wet leaf surface for an extended time, usually about 24 hours. The wet leaf surface allows the fungal spores to swell, germinate and penetrate the plant and the bacteria to swim to a natural opening in the leaf surface, such as a stomate.

Leaf-infecting diseases are more severe if the fungus or bacterium is present at bud break, when the leaves are tender and new.

If the weather is dry during bud break, infection occurs later in wet weather after leaves are expanded. Late infections may be unsightly but seldom *harm the plant.*

Bacterial Leaf Spots

Symptoms: Bacterial leaf spots may appear similar to fungal leaf spots, so it may be difficult to distinguish the two. Leaf spots caused by bacteria are often initially light green and look "water-soaked". Later, these leaf spots usually turn brown or black and may have definite margins.

Cause: Many species of bacteria may cause bacterial leaf spots of landscape plants. These bacteria are often splashed from the soil onto wet foliage, where they enter a leaf through its breathing holes (stomates) or wounds.

Thereafter they spread from leaf to leaf when plants are watered or during rainy periods. Bacterial leaf spot of English Ivy is a good example of a common bacterial leaf spot.

Control: Avoid using high nitrogen fertilizers on susceptible cultivars. Only fungicides containing copper may be effective against bacterial blights and bacterial disease, but copper may burn foliage in some cultivars. The antibiotic streptomycin is used in commercial agriculture, but is not available to the homeowner.

Fungal Leaf Spots

Symptoms: Leaf spots are characterized by brown or black spots randomly scattered across the leaf. The spots may or may not have the appearance of concentric rings. The margins of the spot may be a different colour than the centre of the spot.

Cause: Leaf spots are caused by many different fungi that attack landscape ornamentals. Rainy periods favour their sporulation and spread to unaffected foliage. Infection may occur at any time, but the disease is usually most severe on new, emerging foliage during rainy periods.

Control: Fungi that cause leaf spots form spores on the leaves. The new spring infections are commonly caused by spores from last fall's leaves.

Therefore, rake and burn or compost last season's leaves before buds break in the spring. Help foliage dry quickly after rain, heavy dew, or sprinkler watering.

In the case of deciduous trees and shrubs, prune out overthick growth for better penetration of solar radiation and better air circulation to aid in drying foliage.

In the case of conifers, it is most important to control weeds that surround the lowest branches. Weeds inhibit air circulation and increase humidity, favouring fungal infection of needles.

Fungal Sinusitis Treatment: a Life-long Commitment

Types of Disease

Fungal sinusitis can be invasive or noninvasive, said Richard D. deShazo, MD, director of the Division of Allergy and Immunology at the University of Mississippi Medical Centre in Jackson. In noninvasive disease, fungi appear only in mucus from the sinuses. When the disease is invasive, fungi may penetrate the sinus mucosa, submucosa, blood vessels, and/or bone.

Noninvasive fungal sinusitis is most common in patients with allergic rhinitis, asthma, chronic sinusitis, and nasal polyps. These patients, many of whom have undergone multiple sinus surgeries before diagnosis, may have AFS or sinus mycetoma. Diagnosis hinges on histopathologic visualization of fungal elements. For instance, allergic mucin—an eosinophilic, fungi-containing, peanut butter-like substance—develops in the sinuses of patients with AFS. Although other forms of fungal sinusitis can have a thick or gravel-like exudate, only AFS manifests the characteristic allergic mucin.

Sinus mycetoma may occur in any paranasal sinus and, unlike AFS, is often unilateral. Sinus mycetoma patients may present with symptoms of chronic sinusitis. "Some intermittently blow gravel-like material from their noses," added Dr. deShazo.

Invasive fungal sinusitis, which primarily affects immunocompromised patients, exists in three forms: granulomatous invasive fungal sinusitis, chronic invasive fungal sinusitis, and acute fulminate invasive fungal sinusitis. Granulomatous invasive fungal sinusitis occurs almost exclusively in North Africa; it may produce pressure necrosis of the sinus wall, orbital proptosis, or even brain erosion. Chronic invasive fungal sinusitis is characterized by a low-grade, mixed cellular infiltrate; nasal polyposis; and thick, purulent mucus. Decreased vision, impaired ocular mobility, and proptosis may develop if the infection reaches the orbital apex.

Acute fulminate invasive fungal sinusitis is caused by rapid spread of fungi from the nasal mucosa and sinuses into the orbit, vessels, and brain parenchyma. "Prognosis is poor," said Dr. deShazo, "although with aggressive and early treatment, up to 50% of patients may survive."

All Fungi are Pathogenic

Patients are increasingly becoming "living Petri dishes," said Michael G. Rinaldi, PhD, a clinical mycologist. The increasing prevalence of immunocompromised patients—the result of human immunodeficiency virus infection and of treatments that impair the immune system—has meant a growing incidence of mycotic infections, including fungal sinus infections. The list of fungi that can infect the sinuses is long. "Basically, there are no longer any nonpathogenic fungi," indicated Dr. Rinaldi, director of the Fungus Testing Laboratory and a professor of pathology, medicine, and microbiology at the University of Texas Health Science Centre at San Antonio. "These days, given the right immunocompromised person, virtually any fungus can cause human disease."

The unique case of a 26-year-old man with end-stage human immunodeficiency virus infection and a low CD4 T-cell count provides an illustration. The patient, Dr. Rinaldi recounted, unknowingly inhaled spores from the wild mushrooms he had gathered and prepared for a meal. Several weeks later, the patient presented, complaining of itchiness and a fissure in the roof of his mouth.

His physician discovered the head of a mushroom literally growing inside the patient's head. The mushroom protruded down through the fissure from the sinus cavity. Culture confirmed it was the same type of fungus—*Schizophyllum commune*—that the patient had prepared for the meal.

Aspergillus fumigatus and *A flavus* are much more common causes of fungal sinus infection, and they may be lethal. Under the microscope, *A fumigatus* has a smoky, gray-green appearance, said Dr. Rinaldi. In contrast, *A flavus* is lime green in colour. In addition to *Aspergillus* sp, *Bipolaris spicifera* and *Curvularia lunata* often cause AFS.

Identifying some of these organisms is difficult because they closely resemble numerous other types of fungi, including some molds that may be equally deadly. Physicians should therefore always use a good microbiologist to identify the infecting organism when fungal sinusitis is suspected, Dr. Rinaldi recommended, not only because it is good medicine, but because such identification may help determine the most appropriate therapeutic approach.

High Recurrence Rate

A variety of surgical procedures exist for the treatment of fungal sinusitis. The presence of acute fulminant invasive fungal sinusitis constitutes a surgical emergency. "It requires immediate, radical resection of tissue and application of systemic antifungal agents," emphasized Dr. Kuhn. Without such intervention, the patient has a poor chance of survival. The previous treatment for chronic sinusitis was open sinus surgery, such as frontal sinus obliteration or external frontoethmoidectomy. A major disadvantage of open surgery, however, is that it typically involves removal of large amounts of bone and healthy sinus tissue.

The newer intranasal sinus procedures, such as functional endoscopic sinus surgery, are usually far less damaging. "The purpose of functional endoscopic sinus surgery is to remove the obstruction at the outflow tract of the sinuses while preserving all possible mucous membranes and all possible normal structures," according to Dr. Kuhn.

Regardless of the procedure, recurrent disease is common—usually, as a result of incomplete surgery that leaves obstructions likely to impair mucocilliary clearance. For example, the recurrence rate of chronic sinusitis is 55% when a poorly resected uncinate process obstructs the maxillary sinus ostium; it jumps to 80% when residual infraorbital ethmoid cells produce such an obstruction.

Postoperative disease recurrence is also highly likely with allergic fungal sinusitis. In a study of 24 patients with the disease, eight of the nine who underwent surgery had a recurrence; the ninth patient experienced a recurrence after the study was completed. "So the key take-home message," said Dr. Kuhn, is that "every patient who is operated on and gets no medical treatment afterward will suffer a recurrence. You just have to wait long enough."

The two longest periods between surgery and recurrence were 29 and 34 months, suggesting that fungal sinusitis patients require very long-term follow-up. In practical terms, the extreme likelihood of recurrence means "that I need to follow these patients for at least five years," Dr. Kuhn concluded. He recommended monthly follow-up for all patients during the first year after surgery, then bimonthly follow-up during subsequent years for those who are doing well.

Corticosteriods Reduce Recurrences

Further study is required to determine the optimal pharmacotherapy for fungal sinusitis. However, starting long-term oral corticosteroid therapy immediately after surgery appears to reduce recurrences and disease severity, at least with AFS, said Dr. Schubert. By way of evidence, he described a retrospective study of 67 AFS patients who received standard postoperative treatments (eg, anti-inflammatory nasal sprays, antihistamines, decongestants, and allergen immunotherapy) or standard treatment plus oral corticosteroids.

The corticosteroid group received prednisone; 0.5 mg/kg/d was given for two weeks after surgery, followed by 0.5 mg/kg every other day for about three weeks. The dosage was gradually tapered to between 5 and 7.5 mg every other day by three months, and then 5 mg every other day was given for the remainder of treatment (typically, about a year). If repeat surgery was necessary, this protocol was restarted from the beginning.

At two months' follow-up, most patients in the corticosteroid group had mild or no AFS, whereas patients in the standard treatment group typically had moderate or severe disease, Dr. Schubert said. The results were even better at one year.

The rate of repeat surgery for recurrent disease at 200 days' follow-up also reflected the advantage of corticosteroid use: It was less than 20% in the prednisone group versus about 50% in the standard treatment group. Oral corticosteroids provided significant benefit, stressed Dr. Schubert, even when patients did not complete the entire treatment course.

Serial measurement of total serum immunoglobulin E (IgE) levels has prognostic value, the study also showed. Patients with fungal sinusitis typically exhibit high-titer total and fungal-specific serum IgE levels. "If that level increased 10% or more, the chance of needing recurrent sinus surgery was roughly 50%," Dr. Schubert reported. Furthermore, the total serum IgE level corresponded with the severity of AFS, rising when the disease worsened and falling when it improved.

Dr. Schubert also emphasized the importance of using a "steroid burst" (a temporary increase in the prednisone dose, followed by a rapid taper to its normal level) whenever a patient with AFS has an intercurrent acute upper respiratory tract infection. "Absent that maneuver," he said, "patients with acute upper respiratory tract infections have an increased tendency to relapse." Systemic antifungal drugs have not been found beneficial, possibly because AFS is noninvasive, Dr. Schubert suggested. It is unknown whether the disease responds to topical antifungals, which are problematic anyway because of the difficulty of applying them to sinus tissue.

Organic means of Disease & Insect Control

Dormant Spraying

Dormant spraying means exactly what it says; spraying done when the plant are in dormancy. That means no leaves or active growth should be happening. Dormant spraying should be done immediately after pruning in spring and again in late fall when the leaves drop (usually late Oct.) Spraying dormant oil & lime sulfur when there is active growth will burn & damage leaves so if you snooze you loose in this instance! Your dormant spray will damage that tender new growth, setting your plant back, and possibly resulting in deformed growth.

A thorough dormant-spraying offers you your best chance for control of the various fungal diseases that plague roses: powdery mildew, rust, blackspot, insect pests, even downy mildew-all can be reduced by a careful dormant spraying.

Dormant Oil sprays and Lime-sulfur sprays are available at our nursery, and they are very reasonably priced. These materials are usually formulated for combined use. They may be safely used on roses now, while there is no foliage to be damaged. These oils coat the canes & suffocate the eggs & adult insects so spray each rose plant thoroughly, and spray the ground around it as well. Totally saturate the canes making sure you have not missed a minute crevice! Insects & tiny dust like spores of diseases hide in the tinniest of crevices & will come to life & multiply if you miss them. This application is essential if you are having spidermite problems. Those little unbearable pests are very hard to get rid of & a double dose of dormant oil really works wonders. In the fall & again in the spring giving them a double whammy.

Insecticidal Soap

When using insecticidal soaps, spray the entire plant thoroughly and repeat applications frequently. You may have to spray every few days at first until the infestation is under control. Apply in the morning, evening or on overcast days as not to burn tender new leaves if sprayed in full sun. Read carefully all label instructions. Soaps can burn some skin sensitive people or hurt eyes. If mixed too strong can cause severe damage to tender new growth! Rugosa Rose leaves do not like to be sprayed with anything especially insecticidal soaps! Spray only the parts where insects are a real problem. If possible use rain or soft water as soaps adheres better in water without high minerals content.

Organic Tips

- To prevent spreading fungal diseases from plant to plant, disinfect pruning shears between cuts, dip the blades in a 10%

solution of chlorine bleach(1 part bleach to 9 parts water) or pure alcohol or peroxide Some rose growers submerge their bareroots in the bleach solution or peroxide before planting in spring or fall. Only dip them do not leave them in it for any length of time. This works really quite well! Canes need to be completely coated so dipping is the best! Pheromone traps use specific (sex hormones) attractants to lure insects inside where they get stuck or trapped. They are also used to monitor and alert the grower when insects appear on the scene. Sticky cards are a great help for insects such as whitefly (yellow) and midges (blue cards).

- As unfortunate as it is, sometimes we have to eliminate the source of our problems! Getting rid of the troublemakers although painful, is one way to help rid and curtail diseases; especially if you want to grow organically! Removing the varieties that are susceptible (prone) to diseases will help remove the source of infections and overall disease problems. This happened to us early in our nursery business and changed our outlook on what we grew there after! As we loved and grew some "Soel D'or" and "Persian Yellow" hybrid Rosa foetidas very prone to blackspot and more so in wet Nova Scotia climates! After continual sprayings of sulfur we came to the conclusion of roses that are really prone to disease should not be grown period! Especially if your a nursery that has to keep all its stock in top shape. You just can't win with varieties like these, as beautiful as they may be. Be sure to use dormant oil & sulfur the fall & spring after to eliminate all residues left in beds & other infected roses.
- Spraying with broad-spectrum insecticides (those that are effective on ALL insects) reduce many natural predators that feed on the problem pests also. When you use these products you're wiping out pretty much ALL INSECTS where you have sprayed! This includes MR. Bee to the left! Population increases have been studied & noted on the occurrence of spider mites & thrips and the use of broad-spectrum insecticides on them. Reason, Not only are the natural Enemies of these pests killed, but also newer generations have built up a greater resistance to the insecticide and future applications will be less & less effective.
- A strong jet of water from a garden hose can help control many diseases and insect pests. A direct stream to the tips where

aphids are and the undersides of leaves where spidermites flourish will knock and keep these critter at bay. Aphids are very slow moving insects and the chances of those ones getting back on the plant is almost nil due to their predators on the ground and spidermites absolutely HATE WATER, so they won't hang around where they are constantly subjected to this. Because wet foliage encourages fungal diseases, it is wise to do this early in the day or on a windy day so leaves dry off readily. Never at night, if you can help it.

- Blackspot, mildew & insect problems can be reduced by cleaning up and destroying all infected leaves early in the infestation, (Do NOT compost them) Good fall housekeeping is essential to not carry it on into next season. Clean-up all fallen leaves and leaves still on roses. Again don't compost these burn them if you can.
- Botanical insecticides include rotenone and pyrethrum. Although these insecticides are derived from plants, they are not necessarily safe. In a well-managed garden, these should be used only when other methods have failed. They kill both beneficial as well as pest insects. They should be sprayed only at dusk when fewer honeybees are active; rotenone and pyrethrum are toxic to fish and should not be applied near ponds. Rotenone is an insect stomach poison best applied as a 1% solution in water. Pyrethrum is a nerve poison and paralyzes insects on contact. Nicotine, another botanical insecticide, is best not use it is very highly toxic to mammals.
- Organic sprays and dusts include insecticidal soap and diatomaceous earth. Both are effective against soft-bodied insects such as aphids, spider mites, and some caterpillars. Insecticidal soaps penetrate insect membranes, causing paralysis and starvation. Use soft water when mixing and add isopropyl alcohol (2 Tbsp. per liter). Diatomaceous earth is a dust composed of the fossilized skeletons of microscopic sea organisms. The individual particles are sharp and cut insect bodies, causing them to dehydrate and die. It is applied as a dust, or in a water solution. Homemade sprays of ground up garlic and hot peppers can also effectively repel some insect pests.

Cultural and mechanical practices involve handpicking insect pests when they are few in number and crushing them underfoot, keeping the yard and garden free of decaying refuse, rotating the position of crops each year, and maintaining healthy plants with fertile soil.

Bacillus Thuringiensis, "BT"

1. Bt (Bacillus thuringiensis) The Natural Insecticide Bacillus thuringiensis (Bt) is a naturally occurring, soil borne organism that is used for its ability to control certain insect pests especially larvae of flies and months that can severely damage roses and other plants. Bt works in a natural, environmentally friendly manner. Bacterial agents, like Bt, are effective in controlling insects in the larva stage only. In this stage the larvae is usually a caterpillar or worm and to effectively control the pest, you must get it early at this stage both to eliminate the pest and to avoid too much damage to the leaves of the rose. The Bt is applied to the foliage of plants infested with a leaf or needle-eating larva. It is not as important to completely saturate the leaf surface as with insecticidal soaps as the larva need only eat a portion of the leaf with the bacteria on it to do the trick. You'll usually find these critters on the undersides of leaves and usually the same colour as the leaf as not to attract their predators therefore Bt should be applied to the UNDERSIDES of leaves.
2. Bt is also broken down faster in sunlight. Application to the undersides of the leaf surfaces will prolong Bt's activity. High temperatures do not encourage Bt's breakdown.
3. After Bt spores are ingested by larvae, they grow and reproduce, meanwhile producing crystalline toxins. The crystalline toxins paralyze the digestive tract of the larvae causing it to cease eating. Death will follow, but time of death can range anywhere from 12 hours to 5 days after ingestion This depends on the amount of Bt ingested, the size and variety of the larvae and variety of Bt used for control. There are different strains or varieties of Bt available that have been selected for the control of specific insects. The one we promote for roses is Bt variety kurstaki (BTK). This controls the European corn borer, tomato hornworms, fruitworms, cabbageworm, cabbagelooper, spring and fall cankerworm, spruce budworm, and other caterpillar-like larvae. Other Bt varieties like San Diego (BTSD) controls early larvae of the Colorado potato beetle and Bt variety israelensis (BTI) controls mosquitoes, black flies and fungus gnats.

Bt was used extensively in Nova Scotia's forests these years for control of the Tussock Month as great treat to the evergreen forest and Christmas Tree Industry. The Bt has gained well-earned popularity because of its distinct advantages over other pesticides such as :

(i) Hazards to humans are negligible although inhalation or contact with eyes or open wounds should be avoided.

(ii) Using it on vegetables, Bt can be used right up until harvest, which allows for a longer-term control as compared to other insecticides requiring a waiting period from time of application to time of harvest.

(ii) There is no waiting period from time of application before re-entering the field.

(iii) Ficial or non-target insects are not harmed.

(iv) Insects that ingest the Bt and later die from it are not dangerous to birds or other animals that may feed the dead insect.

(v) Is not known to cause injury to plants on which it has been applied and is not considered harmful to the environment.

(vi) With any substance you should not depend and use only it as insects tend to build up resistancy over time to them and for the most effective use try to use other means also every now and then.

Organic Earwig Control

Earwigs can really be a menace and to know how to effectively control them organically you must first know their life cycle. Here in Canada the earwig over winters as an adult in nests that is made in cracks and crevice, under rock and boards around the yard. The female lays a portion of her eggs in the fall. In the spring she will finish laying her eggs in the nests she has created. In late March to April, depending on how warm it gets, the males leave the nest. This is why early in the year one will see a few large earwigs. The female will remain in the nest and care for the eggs and the hatched young until late June to July. At this time the young earwigs and the females will leave the nests looking for food. This is the time of year when most people find that earwigs are a problem. This then is the time to focus on controlling them.

Try to confirm the life cycle with your area Dept. of Agriculture knows about when the females open up the nests. The most effective way to control earwigs without insecticides is the use of trapping.

Again knowing the insects habit makes the use of traps fairly effective. Earwigs are night creatures. They feed and move about in the dark. During the daytime they will find areas to hide from the light. Any small crevice or container makes great daytime homes. Traps should be set in the evenings and collected in the morning to remove the captured insects. The best types of traps that seem to work well are: corrugated cardboard; Rolled up magazines or newspapers; small cans with openings punctured in the ends.

Make sure holes are large enough to let the earwig in tubing such as bamboo rods. Two pieces of wood with groves cut and taped together. When collecting the traps, make sure they are collected each day in the mornings. Dispose of the earwigs in a container of diesel fuel or gasoline. This will kill the insects quickly. When placing traps, make sure they are placed in and under shrubs and other dark hiding places that the earwigs would like to stay. Place the traps when the nests are opened up later in the summer. People that report failures with traps result from setting the traps too early. They operate the traps and find that no or very few earwigs are caught. They then stop trapping and when June/July comes earwigs by the "hundreds" are found and they think the traps were not successful. "Wrong" just poor timing and not working along with the earwigs life cycle.

Fungal Leaf Spots

Brown spot is a common disease which occurs frequently on young plants. If warm moist weather conditions continue, infection, and defoliation progress from lower to upper leaves. Usually the hot dry weather of midseason arrests disease development, but the disease may become active again near maturity. If conditions favourable for disease development continue into the season or reoccur before maturity, the resulting defoliation can cause serious yield reductions. The pathogen overwinters on infected leaf and stem debris, and the disease is most severe when soybeans are grown continuously in the same field.

Cotyledons, primary leaves, and lower trifoliolate leaves typically show brown spot symptoms. The brown to red lesions vary in size from pinpoint to 4mm wide, but may coalesce and form larger irregular shaped spots. Severely affected leaves turn yellow and drop. Although there are differences in susceptibility of cultivars, no highly resistant cultivars are available. Control is best achieved by rotation and deep plowing to reduce disease incidence.

Downy mildew, one of the most common foliar diseases of soybean, occurs worldwide, but is seldom a serious problem. This disease reduces seed quality and seed size, if extensive defoliation occurs, yields can also be reduced. Disease development is favoured by high humidity and cool temperatures. Symptoms are found on young plants, but the disease does not become widespread in a field until the late vegetative or early reproductive stages. The increased resistance of older leaves and higher temperatures at mid-season stops disease development before extensive damage occurs.

Lesions start as small light green to yellow indefinite spots on the upper leaf surface. As spots enlarge they become yellow-brown to dark brown. A grayish downy tuft of sporangiophores develops on the lower leaf surface during periods of moist weather. Severely infected leaves drop prematurely.

Cultivars with some resistance are available and should be used. However, many races of the pathogen have been identified, and cultivars which are resistant to all known races have not yet been developed.

Powdery mildew occurs sporadically on soybeans. Yield reductions are possible if foliage is affected during pod fill or extensive defoliation occurs. Plants are susceptible at any growth stage, but symptoms are rarely seen in the field until the mid to late reproductive stages. Disease incidence is greater in seasons with cooler than normal temperatures. The first symptom is usually a circular area of white to light-gray, superficial powdery growth on the upper leaf surface. Under favourable conditions, cool humid nights and mild dry days, these areas expand rapidly, coalesce and cover the entire leaf surface. All green plant parts may be covered by the powdery growth. Heavily infected leaves turn yellow, then brown, and drop prematurely.

Powdery mildew is controlled with resistant cultivars. Most determinate soybean cultivars are highly resistant to powdery mildew, while many indeterminate cultivars are susceptible.

Aerial blight is found in several states, but is primarily of economic importance in Louisiana where extensive yield losses have been reported. The fungus survives in the soil and on plant debris, as well as on certain weed hosts. Infection usually begins at flowering during prolonged periods of high humidity and warm temperatures.

Symptoms include leaf spots, leaf blighting, or defoliation. Leaf spots are irregular water-soaked spots that develop reddish-brown margins during drier conditions. Tissue in the centre of lesions falls out giving the plant a ragged appearance. Brown lesions may also form on the petioles, stems, and pods. Severe infection causes pod blight.

Control strategies include the use of good management practices (seed bed preparation and weed control) and applying foliar fungicides. Some cultivars are more tolerant and should be planted in fields with a history of aerial blight. This destructive disease occurs regularly in areas with high humidity. In other areas, the occurrence of frogeye is sporadic, but is more severe in seasons with frequent rainfall. When a high percentage of leaves are heavily infected, serious yield loss is common. The use of resistant cultivars has reduced losses.

Young expanding leaves are extremely susceptible, while fully expanded leaves are more resistant to invasion. However, because visible lesions develop nearly two weeks after invasion, lesions are never seen on young expanding leaves. Leaves that expand during periods of weather unfavourable for invasion remain relatively disease free and result in "layers" of healthy leaves between "layers" of heavily diseased leaves.

The foliar lesions measure up to 5mm in diameter but may coalesce to form larger spots. Young, fully developed lesions have a gray to brown centre with 'a distinct purplish to reddish-brown margin. As lesions age, the centre becomes tan to nearly white and the margin darkens. Lesions also occur on stems and pods.

Crop rotation and plowing under crop residues will reduce disease incidence. Fungicides applied at the late flowering and beginning seed growth stages will protect against frogeye infection. Several races of the pathogen have been identified, but cultivars with resistance to all known races are available. Cercospora leaf blight is a late season disease that is often mistaken for early senescence. Symptoms are first observed at the beginning of seed set. Extensive losses from the disease have been reported in some years. While this fungus also attacks pods and causes purple seed stain, there is no consistent relationship between the occurrence of leaf spot and purple seed stain.

Symptoms are first observed at the beginning of seed set. Small, reddish purple, angular to irregular lesions occur on the upper leaves which become leathery and dark purplish-red with bronze highlights. This discoloration may extend to the upper stems, petioles, and pods. Heavily infected leaves rapidly turn yellow and fall, mimicking senescence. Lower leaves remain green. Although some cultivars possess resistance, control is primarily achieved by application of foliar fungicides starting at early pod set. Alternaria leaf spot has been reported from all soybean growing areas. This fungus is often a secondary invader following mechanical injury, insect damage, or another disease. Altemaria leaf spot is a late season disease that develops on maturing leaves or on younger leaves if injured. If uninjured, most improved determinate cultivars appear to be resistant until leaves start to senesce. Since lesions develop after the onset of yellowing, yield losses are considered minimal. The causal agent appears to be seedborne.

Lesions are nearly circular unless they are restricted by a major vein or coalesce with another lesion. Single lesions range up to 25mm in diameter, have brown concentric rings, and a well defined margin. Control measures are not considered necessary. Target spot occurs regularly in low incidence in many areas. The disease usually occurs too late in the season to cause economic damage. However, the disease is potentially serious on late-maturing cultivars or on susceptible cultivars in seasons of high rainfall. Leaf symptoms are seen prior to maturity during periods of high humidity or when free moisture is present.

Lesions on leaves are reddish-brown, circular, and variable in size up to 15mm in diameter. Spots may also be found on petioles, stems, and pods. The larger lesions often show distinct concentric zones of dead tissue and may have a narrow, indistinct yellow-green halo. Lesions which form along veins may be narrow and elongated. Severe infections may cause premature leaf drop.Most well adapted, high yielding, determinate cultivars have tolerance to target spot.

8

Characteristics: Biology, Ecology, Uses, Cultivation

Mycology is the study of fungi and mycologists are the people who carry out these studies. New research methods have substantially increased knowledge about the fundamental nature of fungi. Much of this research has focused on fungi that cause plant diseases.

Research on edible fungi has concentrated on a small group of species that are commercially cultivated. Wild edible fungi have, until recently, been relatively ignored by science, though amateur mycologists often documented species they found in field studies, mostly in Europe or countries in which Europeans have settled.

There has always, however, been a keen interest in a small group of valuable wild edible fungi that cannot be cultivated. These include the truffles (*Tuber* spp.), *matsutake* (*Tricholoma* spp.) and *porcini* or *cèpes* (*Boletus edulis*). Their biology and ecology have been studied in some detail – a marked contrast to the many other wild edible fungi used around the world.

The consequence of this neglect is that wild edible species used in developing countries are poorly known.

Some information is available from studies of close relatives in temperate regions. *Russula* and *Lactarius* occur around the world, for example, and knowledge of species in Europe can be applied with some caution and caveats to African species. The main problem is naming and recognizing species. Genera and species concepts were originally based on the narrower range of diversity found in temperate regions and these may require fundamental reappraisal as tropical species become better known. This chapter provides a brief introduction to the larger fungi (macrofungi), with special reference to those that are edible.

The use of specialist terms has been avoided where simpler alternatives are available. Field guides contain useful glossaries and there are an increasing number of Web sites that help in understanding technical terms. The *Dictionary of the fungi* is a regularly updated text with details about all fungal genera and other information on mycology.

What are Fungi?

Fungi are a distinct group of organisms more closely related to animals than plants. At present fungi are divided into three separate and distinct kingdoms based on an expanded knowledge of their biochemistry and genetic makeup established especially over the last 30 or so years. It is wrong and misleading to refer to fungi as "plants without chlorophyll".

Despite fundamental differences, fungi are often classified as plants. Understanding the taxonomic status of fungi has little apparent significance to people collecting and selling wild edible fungi, but it is of critical importance in establishing a sound and robust classification system. This ensures that when two people use the same species name they know that they are referring to the same (edible) fungus.

The classification of fungi with plants has inadvertent practical consequences. It is not always clear whether ethnobotanical studies include wild fungi, as is the case with a study from Turkey. Ethnomycology is the correct term that indicates fungi are involved. On a similar track, flora refers only to plants.

The equivalent term for fungi is mycota. These fungal terms may be unfamiliar but their use helps to identify published information on wild edible fungi clearly that may otherwise be ignored or missed.

Structure and Feeding

Fungi come in many shapes, sizes and colours. Macrofungus (plural: macrofungi) is a general category used for species that have a visible (to the unaided eye) structure that produces spores, such as a mushroom or truffle. These visible structures are generically referred to as "fruiting bodies".

Fungi consist of fine threads known as hyphae, which together form a mycelium, as in the mould growing on a piece of fruit or bread. The cap of a mushroom or a bracket fungus also consists of hyphae, densely packed together to form the fruiting body. Specialized hyphae produce spores that are dispersed in a number of ways. They can be viewed en masse by placing the cap of a mushroom on a piece of white paper and covering it with a glass. The colour, form and way in which spores develop help to identify the fungus.

Wild edible fungi are often referred to generically as wild edible "mushrooms". This can be confusing for a number of reasons: edible species have different forms, some with gills and some with pores, some with stems and some without. This book prefers the broader term wild edible fungi to reflect the diversity of forms and also to distinguish them clearly from cultivated mushrooms.

How Fungi Feed

Fungi are dependent on dead and living material for their growth. They obtain their nutrients in three basic ways:

- Saprobic – growing on dead organic matter;
- Symbiotic – growing in association with other organisms;
- Pathogenic Or Parasitic – causing harm to another organism.

The majority of wild edible fungi species are symbiotic and form mycorrhizas with trees. Saprobic edible fungi are also collected from the wild but they are best known and most widely valued in their cultivated forms. Plant pathogenic fungi cause diseases of plants and a small number of these microfungi are eaten in the form of infected host material. The different modes of feeding are shown in Plate 2 and described briefly below.

Saprobic Fungi

Fungi colonize rotting wood and organic matter found in soil. Many species cannot be seen with the naked eye (microfungi) but there are (edible) macrofungi that fruit on fallen logs and bracket fungi that grow from dead or dying parts of standing trees. *Agaricus arvensis* is a commonly collected wild edible species that occurs in pastures and grassy areas. Edible species of *Favolus* are collected from dead wood inside tropical rain forests. The wild edible fungi used by the Yanomam Indians in Brazil are all saprobic and occurred in slash and burn areas where rotting wood was present.

In the wild, the volume and value of saprobic species used as food are small by comparison with the symbiotic edible fungi, though more edible saprobic species are collected. Their overall value is much higher because they are widely cultivated: a recent figure of US$18 billion was quoted for the annual, global trade in cultivated, saprobic species.

Saprobic species need a constant supply of suitable organic matter to sustain production in the wild and this can be a limiting factor in production. *Shi'itake* (*Lentinula edodes*) mushroom cultivation in one area of China is threatened by the supply of suitable tree branches from nearby forests.

Saprobic macrofungi are also highly valued for their medicinal properties. Most are cultivated, though *Ganoderma* spp. (Plate 9) are also collected from the wild. The list of symbiotic macrofungi with medicinal properties is a short one, though there is some indication that they have been studied less because they cannot be cultivated (Reshetnikov, Wasser and Tan, 2001).

Symbiotic Fungi

The most common form of symbiosis associated with wild edible fungi is that known as a mycorrhiza. Many plants depend on these fungus-root associations for healthy growth. A special type known as an ectomycorrhiza (ECM) is found on trees growing in the Taiga in the Russian Federation and the rain forest of Borneo and includes legume trees as well as conifers. Ectomycorrhiza are typically formed by macrofungi and they include many of the key edible species that are collected in the wild, such as chanterelles (*Cantharellus* spp.) and *Amanita* species.

The mycorrhiza helps the tree to grow in nutrient-poor soils, such as the miombo woodland of central and southern Africa. A sheath of hyphae wraps around the root. They penetrate the root structure but not the actual root cells themselves, forming a living contact between the fungus and the tree. The fungus helps the tree gather water from a wider catchment and delivers nutrients from the soil that the tree cannot access. The tree provides the fungus with essential carbohydrates.

Termitomyces contains important wild edible species. These fungi only grow in association with termites and their nests and are dependent on the organic matter brought by the insects from their feeding on trees. Although *Termitomyces* are saprobic, they are symbiotic with termites. Twenty edible species of *Termitomyces* have been recorded from Africa and Asia. They are regularly collected and also sold. *T. titanicus* is the world's largest edible fungus, although other species are much smaller.

Rural people have long associated the appearance of edible fungi with particular trees and have incorporated this in local names. In southern Africa, *chimsuku* and *kamsuku* both describe *Lactarius* spp. that grow under *masuku* trees.

Some edible ectomycorrhizal fungi produce their fruiting bodies underground. The best known examples are the truffles (*Tuber* spp.: Plate 4). Over 400 species of edible ECM have been recorded. There are also many ectomycorrhizal fungi which produce fruiting bodies that are not edible or are poisonous.

The production of fruiting bodies depends on a complex set of factors and in some years production can be negligible. In Botswana, 14 tonnes of *Terfezia pfeilii,* one of the "desert truffles", were bought from one small community in one season; the next year only four fruiting bodies were located over a much larger area.

The lack of certainty of harvests from one year to the next makes it difficult to plan commercial exploitation and some attempts have been made to overcome this by "cultivating" key mycorrhizal species such as *Tricholoma matsutake.* Trees are successfully infected with truffles and managed under controlled conditions in Italy and elsewhere, but the time, effort and money required are only justified – assuming a good knowledge of the ecology of the fungus concerned – for the most valuable edible mycorrhizal species.

Table: *Plant families with edible ectomycorrhizal fungi*

Family	***Examples***
Betulaceae	*Betula* (birches)
Caesalpinioideae	*Afzelia, Brachystegia, Isoberlinia, Julbernardia*
Casuarinaceae	*Casuarina*
Cupressaceae	*Cupressus*
Dipterocarpaceae	*Shorea, Dipterocarpus, Monotes*
Euphorbiaceae	*Uapaca*
Fagaceae	*Castanea* (chestnut), *Castanopsis, Fagus* (northern beech), *Nothofagus* (southern beech), *Quercus* (oak)

Contd...

Family	***Examples***
"Legumes"	*Acacia*
Myrtaceae	*Eucalyptus*
Pinaceae	*Pinus* (pines), *Picea* (spruces), *Abies* (firs), *Larix* (larches)
Papilionoideae	*Pericopsis*
Nyctaginaceae	*Neea*

Tree species can form mycorrhizas with more than one fungus, and a fungus may associate with more than one tree. Some ECM are "native" to a region: in Madagascar an edible *Russula* grows on exotic eucalyptus. Other edible ECM have been introduced and *Boletus edulis* is now found throughout southern Africa following the establishment of pine plantations. ECM have been most intensively studied in the past on temperate tree species but there have also been steady advances on tropical ECM in Africa. Lichens are "self-supporting" associations between fungi and an alga or cyanobacterium and are the final example of a symbiosis that has edible properties. A lichen is a biological and not a systematic group and several valuable species are eaten by people in Europe, Asia and North America and used for other economic purposes. They are not included in this book. Further information is available from a number of sources.

Plant Pathogens and Parasitic Fungi

In several countries people eat plant material infected with plant pathogenic fungi. Maize cobs infected with the smut fungus *Ustilago maydis* are consumed in large quantities in Mexico, both fresh and canned. They are known locally as *huitlacoche* or *cuitlacoche. U. maydis* is a microfungus: it does not form a visible fruiting body and the only signs of its presence are a mass of dark spores. The cobs appear to become sweeter as the result of fungus attack, and similar changes have been noted for the edible rust fungus *Cronartium conigenum* on pines in Mexico. Other examples include: *Ustilago esculenta* on wild rice; *Sporisorium cruenta* on sorghum in China, personal communication: *Eating* Sporisorium cruenta *in China*); winged bean infected by *Synchytrium psophocarpi* in Indonesia.

Hypomyces lactifluorum is a parasite macrofungus that grows on other macrofungi. It is eaten from Canada through to Guatemala and completes the range of ecological niches occupied by wild edible fungi.

Identification

Local and Scientific Names

Local names have been well documented in Mexico, China and can be checked online for Malawi 4 against the equivalent scientific names. Each of these countries has a rich lexicon of names and terms, a sign of the importance of wild edible fungi to rural people. Some local names have been adopted more widely, particularly for valuable edible fungi. *Boletus edulis* is commonly referred to by its French (*cèpe*) or Italian name (*porcino*– plural *porcini*), and *Tricholoma matsutake* by its Japanese name of *matsutake*.

The system of scientific names aims to remove doubt about the fungus being described. A person with *Cantharellus cibarius* in Nepal knows they have the same fungus as someone in Mozambique, assuming both have been accurately identified.

The scientific name or binomial has two parts. The first name is the genus (*Cantharellus*) followed by the species name (*cibarius*). Named varieties exist for some species but their scientific validity is often uncertain.

Local names for edible fungi are based on shape, taste and other properties that are distinctive or important to people. The lichen (*Umbilicaria esculenta*) and an edible fungus (*Auricularia auricula-judae*) have similar common names in Hunan – *Yan-er* (ear of a rock) and *Mu-er* (ear of wood) respectively. This identifies where they grow and can be collected. Mycologists are sometimes wary of local classifications because they are based on scientifically unreliable characters.

Local names provide important clues to the uses and importance of edible fungi to people and there is much to be gained from their study. Local names allow researchers to learn about collecting practices, to analyse markets and to talk with forest managers and others who lack formal training in science and are unfamiliar with genera and species names. Examples of ethnographic studies involving wild useful fungi are list. Guidelines for conducting such studies are available from a number of different sources.

Local and scientific classifi-cations serve two different groups of people and neither is infallible. Edible species of *Boletus* are not eaten in parts of the United Republic of Tanzania, for example, reflecting local custom rather than scientific fact. Field guides often disagree on which species are edible, either because they are cautious about recommending species that require pre-cooking or because the authors are unaware of local customs in different parts of the world.

What is clear, however, is that there are many poorly described species sold and collected for personal use in developing countries. The rate of discovery is directly related to funding for projects and the ability to draw upon mycological expertise from different countries. Work in the United Republic of Tanzania, Mozambique and Malawi, Burundi and Benin emphasizes the richness of the tropical, edible mycota and how much remains to be done. In the absence of such mycological expertise local names can provide useful information, particularly if dried specimens are available for later examination.

An accurately identified specimen with a scientific name for that species ensures that any new knowledge can be reliably used. A scientific name is the most useful way of finding out whether a species is edible or poisonous, or if it has medicinal or other useful properties. An importer does not need to know if the *pied de mouton* from Bulgaria is *Hydnum repandum* since the genus contains only edible species, but an Italian buyer will pay less for the ordinary *Tuber sinosum* from China compared with other more valuable species. In this instance a scientific name reliably and uniquely describes the fungus in question, for which information can be gleaned from the literature.

The scientific names for fungi are constantly changing – an indication of how much there is still to discover about the diversity of species. New names are proposed and generic boundaries adjusted, both as the result of new discoveries and a revision of the relationships between species. When a new species is proposed it is judged against guidelines and rules drawn up and regularly revised by scientists. The correct publication of a new name does not mean that scientists agree on its taxonomic status. The boundaries between genera and species are open to different interpretations and that is why there are "preferred" rather than "correct" scientific names for fungi.

These changes and uncertainties have important practical consequences for people using wild edible fungi. People have to be aware that a species was previously known by a different name or synonym when searching for information: *Termitomyces albuminosus* was once known as *Collybia albuminosa.* Other changes are less dramatic. *Lentinus edodes,* or *shi'itake* now has the preferred name of *Lentinula edodes*. The older "non-preferred" name is still regularly used in publications. Opinions are still divided as to whether *Coriolus* species with medicinal properties should be renamed *Trametes*. *Auricularia auricula-judae,* the "preferred name", appears variously as *Hirneola auricula-judae* and *Auricularia auricula*.

Table 4 lists the preferred names of wild edible fungi that are still commonly referred to by other names. Common spelling mistakes also appear in publications; even minor differences can cast doubt on the identity of a fungus.

The *Dictionary of the fungi* is a standard reference that is regularly revised to list all genera of fungi. Index Fungorum, an Internet resource, allows users to check the preferred or non-preferred status for species names and to find synonyms. This is of considerable practical benefit, although Index Fungorum lacks the backing required to answer fully queries about which scientific name to use for wild edible fungi. This practical need has still to be addressed by the scientific community.

Table: *Preferred (current or "correct") names of economically important wild fungi*

As Published	***Preferred Name***
Armillariella mellea	Armillaria mellea
Auricularia auricula	Auricularia auricula-judae
Xerocomus badius	Boletus badius
Boletus granulatus	Suillus granulatus
Boletus luteus	Suillus luteus
Calvatia gigantea, Lycoperdon gigantea	Langermannia gigantea
Collybia albuminosa	Termitomyces albuminosus
Coriolus hirsutus	Trametes hirsuta
Coriolus versicolor	Trametes versicolor
Dendropolyporus umbellatus	Polyporus umbellatus
Fomitopsis officinalis	Laricifomes officinalis
Grifola umbellatus	Polyporus umbellatus
Hericium erinaceum +	Hericium erinaceus
Hirneola auricula-judae	Auricularia auricula-judae
Hydnum imbricatus	Sarcodon imbricatus
Hypsizygus ulmarium	Lyophyllum ulmarium
Lentinus edodes	Lentinula edodes

Contd...

As Published	*Preferred Name*
Lepiota procera	Macrolepiota procera
Lepiota rhacodes	Macrolepiota rhacodes
Panus rudis	Lentinus strigosus
Pleurotus cornucopiae var. citrinopileatus	Pleurotus citrinopileatus
Pleurotus ferulae	Pleurotus eryngii var. ferulae
Pleurotus olearius	Omphalotus olearius
Pleurotus opuntiae	Pleurotus ostreatus
Pleurotus porrigens	Pleurocybella porrigens
Pleurotus tuber-regium	Lentinus tuber-regium
Poria cocos; Wolfiporia cocos	Wolfiporia extensa
Rozites caperata +	Rozites caperatus
Sparassis radicata	Sparassis crispa
Strobilomyces costatispora	Afroboletus costatisporus
Termitomyces eurrhizus +	Termitomyces eurhizus
Tricholoma gambosa	Calocybe gambosa
Tricholoma lobayensis; T. lobayense	Macrocybe lobayensis
Verpa bohemica	Ptychoverpa bohemica

Identifying Species

The genera of wild edible fungi found in tropical and subtropical climates are broadly similar to those found in the mycota of temperate regions. The species diversity is, however, much greater in developing countries and care must be taken when comparing specimens with the narrower range of species illustrated in the many field guides published in Europe and North America. Edible fungi occur in two major taxonomic groups. The basidiomycetes contain the mushrooms, bracket fungi and boletes; the ascomycetes include truffles and morels. There is no simple test for determining edibility. The scientific literature is the best objective source of advice, but local practices and preferences can also reveal useful information. Empirical evidence is the ultimate indication of whether or not a species is edible.

The classical method for identifying a macrofungus involves a microscopic examination of tissues, spores and sporing structures. This will at least ensure that the genus is identified. Identification of the lesser known tropical species may also require examination of reference collections. Useful visual clues can be obtained from photographs in field guides and there are increasing numbers of Web sites with photographs and written descriptions of species. Information on Mexican NWFP provided by the Secretaría de Medio Ambiente – Recursos Naturales (2002) on the Internet includes wild edible fungi and is an excellent example of an online guide that could be developed for other regions.

Expert identifications can be costly, although paying for an identification does provide a guarantee of getting a response to a query. Preserving specimens is always useful and at its simplest provides a local reference for comparing specimens. Most macrofungi are easily preserved by drying.

There are special drying racks for fungi, but these can also be locally improvised, adopting methods used for drying fruits and other food produce. Dried specimens can, if necessary, be sent at a later date for scientific identification and should be accompanied by field notes and/or colour photographs.

Molecular tools are commonly used to identify plant pathogenic fungi and have also been applied to truffle species in order to detect which species are used in prepared foods. The practical application of these tools for identifying and characterizing edible macrofungi has still to be explored.

Major Groups of Wild Fungi

There are more than 200 genera of macrofungi which contain species of use to people, mostly because of their edible properties. A clear distinction is made in this book between those recorded as simply "edible" and those that are actually eaten ("food"). To include all edible species as "food" would greatly overstate the number of species consumed by people around the world. Wild fungi with medicinal properties are also valued by rural people in several countries, though this is of secondary importance.

The major genera of wild edible fungi are described in Table 5, with brief notes on medicinal species. The genera of wild edible fungi can be divided into two categories: those containing species that are widely consumed and often exported in significant quantities, such as *Boletus* and *Cantharellus*; and those with species that are eaten widely, usually in small amounts, and rarely if ever traded beyond national boundaries.

Medicinal Mushrooms

Medicinal mushrooms are attracting greater scientific and commercial interest, prompted by a renewed awareness of the use of such material in traditional Chinese medicine. The *International Journal of Medicinal Mushrooms* began publication in 1999 and is an important source of information for this expanding field of research.

Ceremonial Aspects

The ceremonial and religious roles played by wild fungi in different cultures are closely associated with hallucinogenic properties. This has attracted much scientific and personal interest, particularly in Mexico. Globally this use of wild fungi is of minor or no relevance to most countries.

Edibility and Poisonous Fungi

Many macrofungi are not worth eating or are simply inedible. This worthless group of species – as defined by their edibility – significantly dwarfs the very small number of toxic or poisonous species, of which there are only a very few that can kill. Yet it is also true that this very small group of lethal species has significantly shaped attitudes to eating wild fungi, creating potential barriers to wider marketing in many places.

Knowing the scientific name of a fungus provides a good indication of its edibility. In some cases the genus alone will suffice; all known *Cantharellus* species are edible (though not equally tasty). On the other hand, *Amanita* contains both exquisite edible and deadly poisonous species. The only reliable guide to edibility is the knowledge that someone has eaten a particular type – and survived. Local practices and preferences are therefore another useful source of information.

There are conflicting reports in field guides about edibility. Some recommend eating species that others reject as poisonous. People from eastern Finland regard the false morel, *Gyromitra esculenta,* as a culinary delicacy once it has been carefully pre-cooked. Guides in the United States and elsewhere state emphatically that the fungus is poisonous and should not be eaten. Other examples of conflicting advice are summarized.

What Species are Eaten?

A total of 1 154 edible and food species have been recorded from 85 countries. The species eaten in one country or region often differ from nearby areas and in some cases there are dramatic changes in tradition. The Mesoamerican tradition of eating wild edible fungi continues from Mexico to west Guatemala then is absent from much of Honduras and Nicaragua, even though both contain forest areas that in theory support production of edible fungi.

The number of species eaten is sometimes only a fraction of those available. Only 15 of the 284 edible species in Armenia are regularly eaten. In two districts of Turkey, 12 out of a possible 29 edible species were collected and eaten. The reasons for these different patterns of use are not always clear but there is a trend of less frequent use as people move away from the land. Rural people in Guatemala have a positive yet informed approach to eating wild fungi which people living in cities lack. Educated people living in towns in Malawi lose the strong local traditions that rural communities maintain and even acquire a suspicious attitude towards wild fungi.

In parts of the United Republic of Tanzania boletes are thought to be poisonous. In Colombia there is no apparent tradition of eating wild fungi in the Andean regions, though they occur widely. *Tricholoma matsutake* was of little local interest in Sichuan, China prior to Japanese demand that stimulated an export trade in the late 1980s and appears to have prompted wider local consumption. A similar event took place in the Pacific northwest, though with *Tricholoma magnivelare* (Redhead, 1997). This was collected and eaten by Japanese settlers in the 1930s (Zeller and Togashi, 1934) but at the time this did not arouse much, if any, local interest.

Poisonous Species

A review of poisoning incidents in official and informal publications shows that the frequency of such events and the effect on humans are overall less than that suggested by attendant publicity (Logemann *et al.*, 1987). During the search for information on wild edible fungi, about 170 poisonous species were noted. Most are either related to edible species or confused with them. There are, of course, real dangers in collecting and consuming poisonous fungi, but these should be seen against the wider background of millions of people collecting and eating wild fungi safely on a regular basis.

Several popular and highly esteemed edible species are poisonous when raw. Few people eat them in this condition and risks of poisoning are in reality small. Poisonous mushrooms vary in their effects from mild stomach and digestive upsets to more serious problems such as liver damage. The solutions to these potential risks include providing local advice on which species to collect and which ones to avoid (Plate 3) and publicity campaigns that highlight potentially poisonous species on posters. Mr Sabiti Fides, a trader in Malawi, took a more direct route by eating mushrooms in front of his customers.

In southern Africa roadside sellers only offer "safe species" and most market places are a reliable means of obtaining known, edible wild fungi. Problems can occur with "contamination" in markets but such incidents are most uncommon.

Finland has trained mushroom advisers covering all rural areas. The *svamp* "police" based in some town centres in Norway help collectors identify edible species, and there are similar schemes in other countries.

Poisonings are associated with a number of events:

- young children collecting indiscriminately and eating raw mushrooms;
- immigrants arriving in a new country and wrongly identifying a local species that turns out to be poisonous;
- food shortages and economic hardship force people to hunt for food;
- different physiological responses to an "edible" fungus.

Mexicans living in California have eaten *Amanita phalloides* – a poisonous species not found at home – thinking it was the edible *Volvariella volvacea.* The guide for edible mushrooms in Israel is written in Hebrew and Russian, following the arrival of over one million Russians in the 1990s and their strong tradition of collecting wild edible fungi. One Russian was poisoned when he too mistook a poisonous species for an edible species known from his home country. Other reports suggest a certain recklessness amongst Russians in choosing which species to collect and eat.

Some people eat *Laetiporus sulphureus* without any ill-effects while others feel ill. The suggested reason is that physiological responses by people differ but there could also be different strains of the fungus, which differ in chemical composition. Little is known about this particular feature for poisonous or potentially poisonous species.

A summary of well-publicized incidents of widespread poisoning. There has been a spectacular rise in poisonings and deaths in Ukraine in the last decade. Various reasons have been given, including a dramatic economic downturn and the desperate search for food or produce to trade in local markets.

Regular reports of poisonings in the United States appear in the journal *McIlvainea*. These incidents are insignificant by comparison with the thousands of people who collect and consume wild fungi without any reported problems. Millions of other people around the world also regularly eat wild edible fungi without any ill-health effects, and it is important to keep a sense of perspective when reviewing the reported incidents of poisoning.

Contamination of Wild Edible Fungi

The Chernobyl accident in Ukraine in the 1980s prompted investigations of radioactive materials in sources of wild food and particularly wild edible fungi. Broader concerns about the accumulation of heavy metals and pollutants by macrofungi have also been expressed.

A study of radiocaesium intake via consumption of wild fungi in the United Kingdom concluded that intake depended more on the species eaten than the weight consumed. Mycorrhizal fungi had a significantly greater radioactivity compared to saprobic or parasitic species. Consumption of wild edible fungi in the United Kingdom is small by comparison with other countries but the study gives a general indication of the potential health risks.

One reported case of contamination concerned the accidental mixing of potentially poisonous wild species with wild edible fungi imported by the United States. Such events are rare, however, and there are no known instances of this causing any damage to human health in Europe.

Cultivation of Edible Fungi

There are nearly a hundred species of fungi that can be cultivated. All are saprobic. Commercial markets are dominated by *Agaricus bisporus*, *Lentinula edodes* and *Pleurotus* spp. and these account for nearly three quarters of the cultivated mushrooms grown around the world. The major cultivated species are grown on a variety of organic substrates, including waste from producing cotton and coffee. The technologies are well established and successful mushroom industries have been established in many countries. There has been a huge increase in production in the last ten years, mostly as a result of increased capacity in China.

Reports from Africa, Mexico and Amazonia in Brazil suggest that mushroom cultivation offers economic opportunities as well as nutritional and health benefits. Small-scale cultivation takes place throughout China and could provide a suitable model for technology transfer.

The cultivation of the paddy straw fungus (*Volvariella volvacea*) is integrated with rice production in Viet Nam. Wherever saprobic species are cultivated they require a steady supply of raw materials. The expansion of *shi'itake* production in Qingyuan, China ("the mushroom capital of the world") led to a serious depletion of local forests that supplied the wood on which to grow this edible fungus.

The number of saprobic species being cultivated is steadily increasing and information and practical advice are readily available. Ectomycorrhizal fungi can also be "cultivated". Trees are inoculated with truffle fungus that must then infect the roots and form the ectomycorrhizae. The trees are carefully tended to encourage production of the truffles. Methods for "cultivating" truffles are constantly being refined and improved.

Fungi

Fungi are one of the five kingdoms of organisms. Like higher plants (of the kingdom Plantae), most fungi are attached to the substrate they grow on. Unlike plants, fungi do not have chlorophyll and are not photosynthetic. Another key difference from plants is that fungi have cell walls composed of chitin, a nitrogen containing carbohydrate. All fungi have nuclei and the nuclei of most species are haploid at most times. Many species have two or more haploid nuclei per cell during most of the life cycle. All fungi reproduce asexually by spore production. Most species reproduce sexually as well.

General Characteristics

The different taxonomic groups of fungi have different levels of cellular organization. Some groups, such as the yeasts, consist of single-celled organisms, which have a single nucleus per cell. Some groups, such as the conjugating fungi, consist of single-celled organisms in which each cell has hundreds or thousands of nuclei. Groups such as the mushrooms, consist of multicellular, filamentous organisms which have one or two nuclei per cell. These multicellular fungi are composed of branched filaments of cells called hyphae. The hyphae, in turn, often mass together to form a tissue called mycelium.

Mycology, the study of fungi, has traditionally included groups such as the cellular slime molds, plasmodial slime molds, water molds, chytrids, and several other groups of fungus-like organisms. Most modern biologists consider these groups as diverse assemblages of organisms unrelated to the true fungi considered here. However, it should be emphasized that biologists are very uncertain about the evolutionary relationships of these other groups and the true fungi.

Nutrition and Ecology

Most species of fungi grow on land and obtain their nutrients from dead organic matter. Some fungi are symbionts or parasites on other organisms. The majority of species feed by secreting enzymes, which partially digest the food extracellularly, and then absorbing the partially digested food to complete digestion internally. As with animals, the major storage carbohydrate of fungi is glycogen. Fungi lack the complex vascular system found in higher plants, so their transport of food and water is less efficient.

Along with bacteria, fungi have an important ecological role in the decomposition of dead plants, animals, and other organic matter. Thus, fungi are ecologically important because they release large amounts of carbon dioxide into the atmosphere and recycle nitrogen and other important nutrients within ecosystems for use by plants and other organisms.

Some fungi are parasitic, in that they obtain their nutrients from a living host organism, a relationship which usually harms the host. Such parasitic fungi usually have specialized tissues called haustoria, which penetrate the host's body. Most of the diseases which afflict agricultural plants are caused by parasitic fungi.

Some examples are corn smut, black stem rust of wheat and barley, and cotton root rot. Some species of fungi can also parasitize animals. Even humans can be parasitized by fungi which cause diseases such as athlete's foot, ringworm, and yeast infections.

Evolution

The Fungi constitute a large and diverse group of organisms. Until the 1960s, fungi were considered members of the plant kingdom. With the advent of the five-kingdom system of biological classification, fungi were assembled into a single kingdom because of their similar ecological roles as primary decomposers of organic matter and their similar anatomical and biochemical features. Recent studies that compare the sequence of amino acids in proteins from fungi, plants, and animals now indicate that fungi share a closer evolutionary relationship to animals than to plants.

The evolutionary ancestry and relationships of the different fungi are not well understood. There are few fossils of fungi, presumably because their relatively soft tissues are not well preserved. There is some fossil evidence that they existed in the Precambrian era (over four billion years ago), although identification of these very early fossils is uncertain.

There is definite fossil evidence for fungi in the lower Devonian (about 400 million years ago) period. Fossils of all the major groups of fungi are found in the Carboniferous period (about 300 million years ago). By the late Tertiary period (about 20 million years ago), the fossil record shows a rapid and divergent evolution of fungi. Many of the Tertiary fossils of fungi are similar to existing species.

In addition to the incomplete fossil record, there are at least two other reasons why the evolutionary relationships of fungi are not well-known: they tend to have simple morphologies and they lack embryos which follow a definite sequence of developmental stages. Biologists often use these two features to determine evolutionary relationships of animals and plants.

Another complication is that many fungi presumably evolved similar morphologies by convergent evolution. That is, unrelated species may share a common morphology because they have been subjected to similar selective pressures.

The relatively new technique of molecular systematics is particularly useful in the study of the evolution of fungi. This technique compares the sequence of DNA segments of different species to determine evolutionary relationships. One important finding from this new technique is that the plasmodial slime molds, cellular slime molds, and water molds are only distantly related to the true fungi (the taxonomic groups considered here).

Evolutionary relationships among organisms also are being studied by comparing the sequences of ribosomal RNAs and transfer RNAs from different organisms. Although these RNAs are similar to DNA, they have structural roles in cells, rather than coding for proteins as do DNA and messenger RNAs.

Therefore, the sequences of ribosomal and transfer RNAs tend to be more conserved through evolution. Comparison of these RNAs among fungi, plants, and animals also suggests that fungi are more closely related to animals than to plants. Future work in molecular systematics is expected to tell us more about the evolution and relationships of the fungi.

Classification

Biologists have estimated that over 200,000 species of fungi exist in nature, although only about 100,000 have been identified so far. Since classification schemes of organisms are usually based on evolutionary relationships, and the evolutionary relationships of fungi are not well known, biologists have proposed numerous classification schemes for fungi over the years. Below, we consider the five major phyla that nearly all mycologists would agree belong in the kingdom of Fungi.

Zygomycota, Conjugating Fungi

Species in this phylum reproduce sexually by forming a zygospore, a thick-walled, diploid cell which contains thousands of nuclei. There are about 600 species in this phylum. Most species are terrestrial and feed on organic matter, although there are a few parasitic species.

The conjugating fungi are coenocytic, in that they have a continuous mycelium, containing hundreds or thousands of haploid nuclei, with no divisions between them. However, the Zygomycota do have septa (cross walls) between their reproductive structures and the rest of their mycelium.

The conjugating fungi have a life cycle that includes a sexual phase and an asexual phase. In the asexual phase, thousands of spores develop inside a sporangium, a small spherical structure. The sporangium grows on the tip of a sporangiophore, a specialized aerial hypha, typically about as thin as a hair.

In the sexual phase of their life cycle, these fungi form specialized hyphae, called gametangia, which are of two different strains (sexes), plus and minus. The plus and minus strains are very similar morphologically, but differ physiologically and biochemically. Plus and minus gametangia conjugate with one another and form a structure with hundreds or thousands of nuclei from each strain.

Then, a thick-walled structure, called the zygospore, develops from the conjugated gametangia. Inside the zygospore, the many thousands of nuclei from the plus and minus strains pair off and fuse together to form thousands of diploid nuclei. The zygospore is typically spherical in shape and has a thick, dark outer wall. It usually remains dormant for several months or more before development continues.

As the zygospore germinates, it produces germsporangia which are born on germsporangiophores, structures morphologically similar to the asexual sporangium and sporangiophore. The germsporangium contains thousands of haploid germspores which arose from the diploid nuclei of the zygospore by meiosis. Each germspore is liberated, germinates, and gives rise to a new haploid mycelium.

One of the best known of the conjugating fungi is *Phycomyces blakesleeanus*, a species which grows on animal feces in nature. The sporangiophores of *Phycomyces* respond to a variety of sensory stimuli. For example, they bend in response to light (phototropism), gravity (gravitropism), wind (anemotropism), and nearby objects (avoidance response). Physiologists and biophysicists have intensively studied the response to light.

One important finding is that the light sensitivity of the sporangiophore is about the same as the eyes of humans. Furthermore, like humans, the sporangiophore can adapt to a one-billion-fold change in ambient light intensity. One of the pigments involved in the extraordinary light responses of *Phycomyces* is a flavin (vitamin B2) bound to a special protein. This pigment is commonly called the blue light photoreceptor, since it is most sensitive to blue light.

Ascomycota, Sac Fungi

Species in this phylum reproduce sexually by forming a spore-filled structure called an ascus, which means An American fly agaric (*Amanita muscaria formosa*). This mushroom is very common in all of North America, but is more slender, tinged with a salmon-like coloration, and somewhat more rare in the southern states. Photograph by Robert J. Huffman. Field Mark Publications. Reproduced by permission. literally "a sac."

The hyphae of the sac fungi are divided by septa with pores, that is, they have perforated walls between adjacent cells. They reproduce asexually by producing spores, called conidia, which are born on specialized erect hyphae, called conidiophores. The sac fungi are typically prolific producers of conidia.

The sac fungi also have a sexual reproduction phase of their life cycles. In the first step of this process, compatible hyphae fuse together by one of several different methods. Second, the nuclei from the different hyphae move together into one cell to form a dikaryon, a cell with two haploid nuclei. Third, several cell divisions occur, resulting in several cells with two different haploid nuclei per cell. Fourth, nuclear fusion of the two haploid nuclei occurs in one of these cells, the ascus mother cell. Fifth, the ascus mother cell develops into an ascus.

Then, meiosis occurs in the diploid cells and, depending on the species, four or eight haploid ascospores form inside the ascus. In some species, such as the fleshy and edible morels, a large number of asci are massed together to form an ascocarp.

This large phylum of fungi includes many species which are beneficial to humans. For example, the yeasts are a major group of ascomycetes. Different yeasts in the genus *Saccharomyces* are employed by bakers, brewers, and vintners to make their bread, beer, or wine. Truffles are subterranean ascomycetes which grow in association with tree roots. Traditionally, pigs have been used to sniff out these underground fungi, so that French chefs could use truffles to complement their finest cuisine.

Some other ascomycetes are significant plant pathogens. For example, *Endothia parasitica* is an ascomycete which causes chestnut blight, a disease which virtually extirpated the American chestnut as a mature forest tree. *Ceratocystis ulmi* is a pathogenic ascomycete which causes Dutch elm disease, a scourge of American elm trees. *Claviceps purpurea*, the ergot fungus, infects agricultural grains, and when ingested can cause intense hallucinations or death due to the presence of LSD (D-Lysergic acid diethylamide).

Another well known ascomycete is *Neurospora crassa*, the red bread mold. The ordered manner in which the eight spores of this fungus align during sexual reproduction allows geneticists to construct a map of the genes on its chromosomes. Earlier in this century, biologists used *Neurospora* as a model organism to investigate some of the basic principles of genetics and heredity.

More recently, biologists have shown that the mycelium of this species can produce spores at approximately 24 hour intervals, a circadian rhythm, in a constant environment. Many biologists are currently using *Neurospora crassa* as a model organism for investigation of circadian rhythms, which occur in a wide diversity of organisms, including humans.

Basidiomycota, Club Fungi

Species in this phylum reproduce sexually by forming spores on top of club-shaped structures called basidia. The club fungi are believed to be closely related to the sac fungi. Both groups have cells which are separated by septa (walls), and both have a dikaryotic phase in their life cycle; a phase with two haploid nuclei per cell.

The septum of the club fungi is somewhat different from those of sac fungi and is referred to as a dolipore septum. The dolipore septum has a bagel-shaped pore in its centre. The club fungi reproduce asexually by producing asexual spores or by fragmentation of mycelium.

The sexual reproduction phase of the club fungi involves three developmental stages of the mycelium. In the primary stage, a haploid spore germinates and grows a germ tube, which develops into mycelium. The mycelium initially contains a single haploid nucleus. Then, its haploid nucleus divides and septa form between the nuclei.

A secondary mycelium forms upon conjugation of two sexually compatible hyphae. The secondary mycelium is dikaryotic, in that it has two haploid nuclei, one from each parent. As the dikaryotic mycelium grows, the cells divide and more septa are formed between the new cells.

Each of the new cells in the secondary mycelium has one haploid nucleus from each parent. This is assured by clamp connections, specialized structures unique to the club fungi. These are loop-like hyphae which connect the cytoplasm of adjacent cells and through which nuclei move during cell division.

In particular, during cell division, one nucleus divides directly into the newly formed cell; the other nucleus divides inside the clamp connection and the two daughter nuclei migrate through the clamp connection in opposite directions to the two daughter cells.

The tertiary mycelium is simply an organized mass of secondary mycelium. It is a morphologically complex tissue and forms structures such as the typically mushroom-shaped basidiocarps commonly seen in nature.

Sexual reproduction of the club fungi begins upon fusion of two primary hyphae to form a club-shaped structure, known as a basidium. Second, the two haploid nuclei inside the basidium fuse together to form a diploid zygote. Third, the zygote undergoes meiosis to form two haploid nuclei. Fourth, these two haploid nuclei undergo mitosis to form a total of four haploid nuclei. These four nuclei then migrate into projections, which form on the tip of the basidium. These projections then develop into four separate haploid spores, each with a single nucleus. In the species of club fungi which are large and fleshy, such as the mushrooms, a mass of basidia form a structure called a basidiocarp. The spores on the basidia are released from the underside of the fleshy gills of the mushroom. The colour and shape of the basidiocarp, as well as the colour of the spores are often diagnostic for species identification.

This large phylum includes species which are known as mushrooms, toadstools, earthstars, stinkhorns, puffballs, jelly fungi, coral fungi, and many other interesting common names. Some species, such as the rusts and smuts, are pathogens which attack agricultural grains. Other species, such as the fly agaric (*Agaricus muscaria*) and some species in the genus *Psilocybe,* produce chemical hallucinogens and have been used by numerous cultures in their religious ceremonies. Another species, *Agaricus bisporus*, is the common edible mushroom found in supermarkets.

An important aspect of the club fungi is the great diversity of alkaloids and other toxic and psychogenic chemicals produced by some species. For example, *Amanita virosa,* a mushroom colloquially known as "death angel," is so deadly poisonous that a small bite can kill a person. A related mushroom is *Amanita muscaria,* known as "fly agaric," which is hallucinogenic. Over the millennia, numerous cultures have eaten the fly agaric as part of their religious ceremonies. For example, R. Gordon Wasson has shown that *Amanita muscaria* is the hallucinogenic plant referred to as "Soma" throughout *Rg Veda,* the ancient religious text. According to *Rg Veda,* the ancient Aryans who invaded India about four millennia ago ingested "Soma" as a euphoriant.

While mushrooms are the best-known club fungi, many other club fungi grow underground as mycorrhizae. Mycorrhizae result from a symbiosis between a plant root and a fungus. In mycorrhizae, the fungus typically supplies nitrogen-containing compounds to the plant, and the plant supplies carbohydrates and other organic compounds to the fungus. Mycorrhizae are very important for the growth of orchids. One reason many orchids are difficult to grow is because they require particular fungal species to form mycorrhizae on their roots.

A recent report investigated a subterranean club fungus, *Armillaria bulbosa*, which is a pathogen on tree roots. The investigators used molecular biology techniques to demonstrate that a single subterranean "indi vidual" of this species in Northern Michigan was spread out over 37 acres (15 ha) and weighed an estimated 22,000 lb (10,000 kg). Based on the estimated growth rate of this species, of about 0.7 ft (0.2 m) per year, this individual was about 1,500 years old.

Deuteromycota, Imperfect Fungi

The Deuteromycota is a heterogeneous group of unrelated species in which sexual reproduction has never been observed. Since mycologists refer to the "perfect phase" of a life cycle as the phase in which sexual reproduction occurs, these fungi are often referred to as imperfect fungi.

These fungi may have lost their sexual phase through the course of evolution. Alternatively, biologists simply may not have found the appropriate environmental conditions to observe development of the sexual phase of their life cycle.

The Deuteromycota are classified as fungi for two main reasons. First, their multicellular tissue is similar to the hyphae of sac fungi and club fungi. Second, they have erect hyphae with asexual spores, called conidiophores, which are similar to those of the sac fungi and club fungi.

Most imperfect fungi are believed to be related to the sac fungi because their conidiophores closely resemble those produced by the sac fungi during their sexual phase. The imperfect fungi are not placed in the Ascomycota phylum because classification of that group is based on the morphology of sexual structures which the Deuteromycota do not have.

The best known fungus in this phylum is *Penicillium*. Some species in this genus appear as pathogenic, blue-green molds on fruits, vegetables, and cheeses. Several other species are important for the making of cheeses, such as blue cheese, Roquefort, and Camembert. Certainly the best known product from this genus is penicillin, the first widely-used antibiotic. Penicillin was first discovered in *Penicillium notatum* over 50 years ago, but is now known to be produced by many other species in this genus.

Mycophycophyta, Lichens

A lichen is a symbiotic relationship between a fungus and an alga, or between a fungus and a photosynthetic cyanobacterium. They constitute a very diverse and polyphyletic group of organisms and are classified together simply because they all result from a fungus-alga symbiosis. In most lichens, the fungal species is in the Ascomycota phylum and the photosynthetic species is a green alga from the Chlorophyta phylum. Typically, the photosynthetic species supplies carbohydrates to the fungus and the fungus supplies nitrogen and other nutrients to the alga. The morphology of a lichen differs from its component species.

Lichens can reproduce by several methods. The fungal component of the lichen can produce spores which are dispersed, germinate, and then recombine with the algal component. Alternatively, the lichen can produce soredia, specialized reproductive and dispersal structures in which the algal component is engulfed by fungal mycelium. Typically, the soredia break off from the thallus, the main body of the lichen.

Ecologists have shown that many species of lichens are very sensitive to air pollutants, such as sulfur dioxide. Thus, they are often used as indicator species for air pollution; the presence of certain lichen species correlates with the cleanliness of the air.

Many lichens can inhabit harsh environments and withstand prolonged periods of desiccation. In the temperate region of North America, lichens often grow on tree trunks and bare rocks and soil. In the arctic and antarctic regions, lichens constitute a large proportion of the ecosystem biomass. Many lichens are even found growing upon and within rocks in Antarctica. In the arctic region, the lichen species known colloquially as reindeer mosses (*Cladonia rangifera* and several other species) are an important food for caribou and reindeer.

Studies of the symbiotic nature of lichens in the late 1800s laid an important foundation for development of the theory of symbiogenesis. This theory says that new life forms can evolve from the symbiotic relationship of two or more independent species. Nearly all modern biologists now agree that symbiogenesis of different bacteria led to the origin of eukaryotic cells, which contain many different organelles, intracellular" small organs" which are specialized for different functions.

9

Principles of Diagnosis

The International Workshop on Principles of Diagnosis is an annual event that started in 1989 rooted in the Artificial Intelligence (AI) community.

Its current focus is broader covering of a variety of theories, principles, and computational techniques for diagnosis, monitoring, testing, reconfiguration, fault-adaptive control, and repair of complex systems. Application of these theories, principles, and techniques to industry-related disciplines and other domains is amongst the important topics of the workshop.

Like the previous workshops in this series, DX-07 encourages the interactions and the exchange of theories, techniques, applications, and experiences amongst researchers and practitioners from different backgrounds: Artificial Intelligence, Control Theory, Systems Engineering, Software Engineering and other related areas, who share an interest in different aspects of diagnosis, and the related fields of testing, reconfiguration, maintenance, prognosis, and fault-adaptive control.

DX is a lively forum that has traditionally adopted a single-track program with a limited number of participants in order to promote detailed technical exchange and debate while at the same time making efforts to develop synergistic approaches to solving real-world problems. We welcome papers on topics that are related but not limited to the following:

- Formal theories and computational methods for diagnosis, that include monitoring, detection and isolation, testing, repair and therapy, reconfiguration, fault tolerance, diagnosability analysis, and other related topics.

- Modeling for diagnosis that includes symbolic, numeric, discrete, discrete-event, continuous, hybrid, probabilistic, functional, behavioral, qualitative, abstractions, and approximation methods. Effective modeling approaches for large systems are of particular relevance.
- Computational issues that address combinatorial explosion, use of structural and hierarchical knowledge, focusing strategies, resource-bounded reasoning, real time analysis, and other related topics.
- Diagnosis processes that include strategies for measurement selection, sensor placement, test actions design, active testing, embedded diagnosis systems, preventive diagnosis, fault tolerance strategies, fault-adaptive control, and distributed diagnosis.
- Bridge between DX (AI-based diagnosis methods) and other diagnosis methodologies: FDI, control-based techniques, statistical and probabilistic methods, design, model checking, machine learning, non-monotonic reasoning, planning, execution, real-time languages, software verification and validation, debugging, and hardware testing.
- Real-world applications and integrated systems in a wide range of fields including transportation systems, space and aeronautics, process industries, medical domains, and bioinformatics. Case studies of tech transfer that resulted in success or failure are especially welcome.

Eating Disorders in Adolescents: Principles of Diagnosis and Treatment

Reaffirmed February 2009

Eating disorders are complex illnesses that affect adolescents with increasing frequency. They rank as the third most common chronic illness in adolescent females, with an incidence of up to 5%, a rate that has increased dramatically over the past three decades. Two major subgroups of the disorders are recognized: a restrictive form, in which food intake is severely limited (anorexia nervosa), and a bulimic form, in which binge eating episodes are followed by attempts to minimize the effects of overeating via vomiting, catharsis, exercise or fasting (bulimia nervosa). Both anorexia nervosa and bulimia nervosa can be associated with serious biological, psychological and sociological morbidity, and significant mortality.

Although eating disorders occur most frequently in adolescents, reports in the scientific literature often combine findings from adolescents with those from adults or report exclusively on adult samples. Unique features of adolescents and the developmental process of adolescence are often critical considerations in determining the diagnosis, treatment or outcome of eating disorders.

Consequently, adolescents need to be considered separately and differentiated from adult patients with eating disorders. This position statement addresses the key issues distilled from the scientific literature and represents a consensus of numerous specialists in adolescent medicine regarding the diagnosis and management of adolescents with eating disorders.

Diagnosis

Diagnostic criteria for eating disorders such as described in DSM-IV may not be entirely applicable to adolescents. The wide variability in the rate, timing and magnitude of both height and weight gain during normal puberty; the absence of menstrual periods in early puberty along with the unpredictability of menses soon after menarche; and the lack of psychological awareness regarding abstract concepts (such as self-concept, motivation to lose weight or affective states) owing to normative cognitive development limit the application of those formal diagnostic criteria to adolescents.

In addition, clinical features such as pubertal delay, growth retardation or the impairment of bone mineral acquisition may occur at subclinical levels of eating disorders. The use of strict criteria may preclude the recognition of eating disorders in their early stages and subclinical form (a prerequisite for primary or secondary prevention), and may exclude some adolescents with significantly abnormal eating attitudes and behaviours, such as those who vomit or take laxatives regularly but do not binge. Finally, abnormal eating habits may result in significant impairment in health, even in the absence of fulfilment of formal criteria for an eating disorder. For all of these reasons, it is essential to diagnose eating disorders in adolescents in the context of the multiple and varied aspects of normal pubertal growth, adolescent development and the eventual attainment of a healthy adulthood rather than by merely applying formalized criteria.

Position: In clinical practice, the diagnosis of an eating disorder should be considered in an adolescent patient who engages in potentially unhealthy weight control practices and/or demonstrates obsessive thinking about food, weight, shape or exercise and not only in one who meets established diagnostic criteria. In such adolescents, an eating disorder should be considered if the teenager fails to attain or maintain a healthy weight, height, body composition or stage of sexual maturation for sex and age.

Medical Complications

No organ system is spared the effects of eating disorders. Although the physical signs and symptoms occurring in a patient are primarily related to the weight control behaviours practised, the health care professional must consider their frequency, intensity and duration, as well as the biological vulnerability conferred by the sexual maturity of the patient. The majority of physical complications in adolescents with an eating disorder appear to improve with nutritional rehabilitation and recovery from the eating disorder, but some may be potentially irreversible. The long term consequences are still to be elucidated.

Medical complications in adolescents that are potentially irreversible include growth retardation if the disorder occurs before closure of the epiphyses; pubertal delay or arrest; and impaired acquisition of peak bone mass during the second decade of life, increasing the risk of osteoporosis in adulthood. These features emphasize the importance of medical management and ongoing monitoring by physicians who understand normal adolescent growth and development.

Just as we endorse early recognition of eating disorders through the use of broad developmentally appropriate criteria, we also endorse early intervention to prevent, limit or ameliorate medical complications, some of which are life-threatening. Adolescents who restrict food intake, vomit, purge or binge in any combination, with or without severe weight loss, require treatment even if they do not meet strict criteria for an eating disorder.

Position: Because of the potentially irreversible effects of an eating disorder on physical and emotional growth and development in adolescents, because of the risk of death and because of evidence suggesting improved outcome with early treatment, the threshold for intervention in adolescents should be lower than in adults. Ongoing medical monitoring should continue until the adolescent has demonstrated a return to both medical and psychological health.

Nutritional Disturbances

Nutritional disturbances are a hallmark of eating disorders and are related to the severity and duration of dysfunctional dietary habits. Although abnormalities of minerals, vitamins and trace elements can occur, they generally are not clinically recognized. Deprivation of energy (calories) and protein on the other hand are especially important to identify because these elements are crucial to growth. Moreover, there is evidence that adolescents with eating disorders may be losing critical tissue components, such as muscle mass, body fat and bone mineral, during a phase of growth when dramatic increases in these elements should be occurring. Complete and ongoing assessment of nutritional status is the basis of management of nutritional disturbances in adolescents with eating disorders.

Position: The evaluation and ongoing management of nutritional disturbances in adolescents with eating disorders should take into account the specific nutritional requirements of patients in the context of pubertal development and activity level.

Psychosocial Disturbances

Eating disorders that develop during adolescence interfere with adjustment to pubertal development and the accomplishment of the developmental tasks necessary to become a healthy functioning adult. Social isolation and family conflicts arise at a time when families and peers ought to provide a milieu to support development.

Issues related to self-concept, self-esteem, autonomy, separation from the family, the capacity for intimacy, affective disorders (eg, depression and anxiety) and substance abuse should be addressed in a developmentally appropriate manner. All patients should be evaluated for co-morbid psychiatric illness, including disorders of anxiety, depression, dissociation and behaviour. Because adolescents usually live at home or interact with their families on a daily basis, the role of the family should be explored during both evaluation and treatment.

Position: All adolescents with an eating disorder should be evaluated for co-morbid psychiatric illness. Mental health intervention for adolescents with eating disorders should address not only the psychopathology characteristics of eating disorders, but also the accomplishment of the developmental tasks of adolescence and the specific psychosocial issues central to this age group. For most adolescents, family therapy should be considered as an important part of treatment.

Treatment Guidelines

Because of the complex biopsychosocial aspects of eating disorders in adolescents, the assessment and ongoing management of these conditions appear to be optimal with an interdisciplinary team consisting of professionals from medical, nursing, nutritional and mental health disciplines. Physical and occupational therapy may be useful adjuncts to treatment. Health care providers should have specific experience in treating eating disorders as well as expertise in working with adolescents and their families. They should be knowledgeable about normal adolescent physical and emotional development.

Both in-patient and out-patient treatments need to be available to adolescents with eating disorders. Factors that would justify in-patient treatment include significant malnutrition, physiological or physical evidence of medical compromise (such as vital sign instability, dehydration or electrolyte disturbances) even in the absence of significant weight loss, arrested growth and development, failure of out-patient treatment, acute food refusal, uncontrollable binging, vomiting or purging, family dysfunction that prevents effective treatment, and acute medical or psychiatric emergencies. The goals of treatment are the same in a medical or psychiatric in-patient unit, a day program or out-patient setting: to help the adolescent achieve and maintain both physical and psychological health.

The expertise and dedication of the members of a treatment team who work specifically with adolescents and their families are more important than the particular setting. In fact, traditional settings such as a general psychiatric ward may be less appropriate than an adolescent medical unit, if one of the latter is available. Some evidence suggests that the outcome for patients treated in adolescent medicine units (both out-patient and in-patient) may be better than that of those treated in traditional psychiatric settings with adult patients.

Smooth transition from in-patient to out-patient care can be facilitated by an interdisciplinary team that provides continuity of care in a comprehensive, coordinated, developmentally oriented manner. Health care specialists with an interest in adolescents are familiar with working not only with the patient, but also with the family, school, coaches and other agencies or individuals who are important influences on healthy adolescent development. Given the evidence that eating disorders can be associated with relapse, recurrence, crossover and the later development of other psychiatric disorders, treatment should be of sufficient frequency, intensity and duration to provide effective intervention.

Position: Adolescents with eating disorders require evaluation and treatment focused on biological, psychological and social features of these complex, chronic health conditions. Assessment and ongoing management should be interdisciplinary and is best accomplished by a team consisting of medical, nursing, nutritional and mental health disciplines. Treatment should be provided by health care providers who have expertise in managing adolescent patients with eating disorders and are knowledgeable about normal adolescent physical and psychological development.

Hospitalization of an adolescent with an eating disorder is necessary in the presence of malnutrition, clinical evidence of medical or psychiatric decompensation or failure of out-patient treatment. Ongoing treatment should be delivered with appropriate frequency, intensity and duration.

Barriers to Care

Interdisciplinary treatment of eating disorders can be time-consuming, relatively prolonged and extremely costly. Lack of access to appropriate interdisciplinary teams or insufficient treatment can result in chronicity, social or psychiatric morbidity, and even death. Some provincial plans limit access to private care resources such as nutrition visits or mental health visits. Absent or low reimbursement rates for psychosocial services results in fewer qualified persons being willing to care for teenagers and young adults with eating disorders.

Some older adolescents are no longer eligible for treatment or coverage because of provincial medical insurance rules. Thus, withdrawal from treatment can occur at an age when leaving home, unemployment or temporary employment is the norm. Some institutions have age limit policies that negatively affect treatment and limit access to care during the transition from paediatric to adult care. Legislation should provide reimbursement for intervention by multiple disciplines for adolescents with eating disorders.

Coverage should ensure that for adolescents, treatment should be dictated by the severity and range of the clinical situation. The promotion of size acceptance and healthy lifestyles, introduction of prevention programs for high risk adolescents, and strategies for early diagnosis and intervention should be encouraged.

Position: Health care reforms should include provisions that address the needs of adolescents with eating disorders and ensure that they not be denied access to care because of absent or inadequate health care coverage.

Critical Care Medicine: Principles of Diagnosis and Management in the Adult

The specialty of Critical Care Medicine has matured from its early stage in which the first textbooks attempted to define a curriculum and establish the legitimacy of a new discipline to its current reality. To affirm this status, Parrillo and Dellinger have assembled an international panel of adult critical care experts and have created a clinically oriented, multidisciplinary and multispecialty review of the subject "aimed primarily at practicing critical care physicians who spend the better part of their time caring for patients in an ICU. Thus the book would be appropriate for critical care internists as well as for surgical or anesthesia critical care specialists. The goal was to produce the acknowledged "best practice" standard in critical care medicine."

The book is organized into 9 parts incorporating 83 chapters that cover not only the wide range of diseases and general procedures that constitute the generally accepted areas of critical care knowledge, but also discuss the other significant issues that confront the practicing physician dealing with end of life, limitation of care, and triage considerations. Administrative considerations are discussed, and the burgeoning responsibilities of critical care practitioners are mentioned. The book addresses the core knowledge requirements through the common technique of addressing disease pathophysiology followed by logical management procedures and best practice guidelines. The parts listed with their integrated chapters create an impressive core curriculum for adult critical care medicine:

I. Critical Care Procedures, Monitoring, and Pharmacology
II. Critical Care Cardiovascular Disease
III. Critical Care Pulmonary Disease
IV. Critical Care Infectious Disease
V. Renal Disease and Metabolic Disorders in the Critically Ill
VI. Neurologic Disease in the Critically Ill

VII. Physical and Toxic Injury in the Critically Ill

VIII. Administrative, Ethical, and Psychological Issues in Care of the Critically Ill

IX. Other Critical Care Disorders and Issues in Care of the Critically Ill

Previous editions of this text have focused heavily on shock states and resuscitation; these subjects are well covered in this edition, and the expertise of the authors is easily appreciated in Part II, Critical Care Cardiovascular Disease. Chapters on shock, heart disease and specific cardiovascular emergencies are well prepared and informative. The mainstay of a critical care textbook rests on its ability to discuss and provide appropriate instruction in the resuscitation and maintenance of patients requiring mechanical ventilator assistance and other organ support.

Complex procedures are well described, but despite the comprehensive nature of this text, practitioners will need to refer to the original citations and subject-specific standard texts, rather than utilizing the descriptions herein as being comprehensive. Certain concepts are ignored; the acid-base chapter is well written and detailed, but space constraints reduce the Stewart discussion of strong ion difference to a paragraph and temperature correction to a sentence. However, the authors obviously recognize the fact and, in keeping with the authoritative nature of the text, provide references that are relevant, comprehensive, and up to date.

Infectious diseases and specific consideration of nosocomial infections are presented, and current prophylactic measures to reduce/prevent such concerns are discussed in detail. Sections on ventilator-acquired pneumonia and catheter-related bloodstream infections are well handled. The section titled “Antibiotic Stewardship” is a good starting point for the remainder of this excellent discussion.

It is important to note that the text incorporates the increasing clinical utilization of ultrasound and echocardiography with chapters detailing critical care applications of transthoracic and transesophageal echocardiography techniques and enhanced diagnostic approaches. Surprisingly, other uses of ultrasound (central venous and arterial catheterization, diagnostic focused assessment with sonography in trauma [FAST] exams, diagnosis of pneumothorax, sinus evaluation, etc.) are not mentioned, and it is hoped that future editions will add this component as the use of bedside sonography continues to expand.

Consideration of transcranial Doppler studies is undertaken in three areas: diagnosis of brain death, assessment of cerebral blood flow, and coma. Obviously, extensive supervised and mentored clinical experience is required before any of these techniques can be utilized independently.

There is an attempt made to discuss the importance of newer educational concepts, but it is interesting to note that "evidence" and "outcome" are not found in the index. Conversely, good discussions on quality improvement in numerous contexts are apparent, and a brief section on palliative and end-of-life care is presented with useful check lists for assessing patients' needs and, perhaps more importantly, those of their families.

However, given the increasing importance of this aspect of critical care and the recognition that it is often the intensivist who first broaches this subject to families, it is anticipated that later editions of this text will pay increasing attention to this important area. A brief introduction to the European-based competency assessment (competency-based training program in ICM (intensive care medicine) for Europe; CoBaTriCE) describes the importance of competency-based evaluations in critical care education. This is consistent with the current focus of the ACGME in defining the 6 core competencies in American Residency education programs.

National quality and/or surgical quality improvement activities are not discussed, and a section on leadership and management skills would be helpful. Increasingly, the critical care physician is asked to manage the important components of systems-based practice that provide the mainstay of current ICU management. Despite the scientific approach to disease management, the attention to clinical pharmacology and physiology and the attention paid to machine and information technology interfaces, the book is lacking in a chapter of nursing care and the importance of this integral interaction especially relevant in sections on sedation and ventilator management.

The editors are true to their Critical Care heritage and have assembled a team of clinician leaders who have created what must be considered the standard textbook for the specialty. It is important to note that Parrillo is the Editor-in-Chief of *Critical Care Medicine* and that he and Dellinger are former Presidents of the Society of Critical Care Medicine. Their credentials to edit such a text are impeccable, and their prestige is evident in the list of contributing authors who joined to create the textbook. As they state in the preface, "our view of critical care medicine is mirrored in the organization of the textbook. Modern critical care is a multidisciplinary specialty that includes much of the knowledge and technology contained in many disciplines represented by the classic organ-based subspecialties of medicine as well as the specialties of surgery and anesthesiology.... Each chapter is designed to provide a comprehensive review ofpertinent clinical, diagnostic and management issues.

This isprimarily a clinical text, thus the emphasis on considerations important to the practicing critical care physician; also presented, however, are the scientific (physiologic, biochemical and molecular biologic) data pertinent to the pathophysiology and management issues." As in all such endeavors, there are missing pieces and some areas are slighted. However, the question for potential readers is whether the editors have fulfilled their mandate to create a comprehensive, articulate, clinically relevant and authoritative work. In this they have succeeded.

Principles of Diagnosis

General Concepts

The clinical presentation of an infectious disease reflects the interaction between the host and the microorganism. This interaction is affected by the host immune status and microbial virulence factors. Signs and symptoms vary according to the site and severity of infection. Diagnosis requires a composite of information, including history, physical examination, radiographic findings, and laboratory data.

Microbial Causes of Infection

Infections may be caused by bacteria, viruses, fungi, and parasites. The pathogen may be exogenous (acquired from environmental or animal sources or from other persons) or endogenous (from the normal flora).

Specimen Selection, Collection, and Processing

Specimens are selected on the basis of signs and symptoms, should be representative of the disease process, and should be collected before administration of antimicrobial agents. The specimen amount and the rapidity of transport to the laboratory influence the test results.

Microbiologic Examination

Direct Examination and Techniques: Direct examination of specimens reveals gross pathology. Microscopy may identify microorganisms. Immunofluorescence, immuno-peroxidase staining, and other immunoassays may detect specific microbial antigens. Genetic probes identify genus- or species-specific DNA or RNA sequences.

Culture: Isolation of infectious agents frequently requires specialized media. Nonselective (noninhibitory) media permit the growth of many microorganisms. Selective media contain inhibitory substances that permit the isolation of specific types of microorganisms.

Microbial Identification: Colony and cellular morphology may permit preliminary identification. Growth characteristics under various conditions, utilization of carbohydrates and other substrates, enzymatic activity, immunoassays, and genetic probes are also used.

Serodiagnosis: A high or rising titer of specific IgG antibodies or the presence of specific IgM antibodies may suggest or confirm a diagnosis.

Antimicrobial Susceptibility: Microorganisms, particularly bacteria, are tested in vitro to determine whether they are susceptible to antimicrobial agents. Some infectious diseases are distinctive enough to be identified clinically. Most pathogens, however, can cause a wide spectrum of clinical syndromes in humans. Conversely, a single clinical syndrome may result from infection with any one of many pathogens. Influenza virus infection, for example, causes a wide variety of respiratory syndromes that cannot be distinguished clinically from those caused by streptococci, mycoplasmas, or more than 100 other viruses.

Most often, therefore, it is necessary to use microbiologic laboratory methods to identify a specific etiologic agent. Diagnostic medical microbiology is the discipline that identifies etiologic agents of disease. The job of the clinical microbiology laboratory is to test specimens from patients for microorganisms that are, or may be, a cause of the illness and to provide information (when appropriate) about the in vitro activity of antimicrobial drugs against the microorganisms identified. The staff of a clinical microbiology laboratory should be qualified to advise the physician as well as process specimens. The physician should supply salient information about the patient, such as age and sex, tentative diagnosis or details of the clinical syndrome, date of onset, significant exposures, prior antibiotic therapy, immunologic status, and underlying conditions. The clinical microbiologist participates in decisions regarding the microbiologic diagnostic studies to be performed, the type and timing of specimens to be collected, and the conditions for their transportation and storage.

Above all, the clinical microbiology laboratory, whenever appropriate, should provide an interpretation of laboratory results.

Manifestations of Infection

The manifestations of an infection depend on many factors, including the site of acquisition or entry of the microorganism; organ or system tropisms of the microorganism; microbial virulence; the age, sex, and immunologic status of the patient; underlying diseases or conditions; and the presence of implanted prosthetic devices or materials. The signs and symptoms of infection may be localized, or they may be systemic, with fever, chills, and hypotension. In some instances the manifestations of an infection are sufficiently characteristic to suggest the diagnosis; however, they are often nonspecific.

Specimen Selection, Collection and Processing

Specimens selected for microbiologic examination should reflect the disease process and be collected in sufficient quantity to allow complete microbiologic examination. The number of microorganisms per milliliter of a body fluid or per gram of tissue is highly variable, ranging from less than 1 to 10^8 or 10^{10} colony-forming units (CFU). Swabs, although popular for specimen collection, frequently yield too small a specimen for accurate microbiologic examination and should be used only to collect material from the skin and mucous membranes.

Because skin and mucous membranes have a large and diverse indigenous flora, every effort must be made to minimize specimen contamination during collection. Contamination may be avoided by various means. The skin can be disinfected before aspirating or incising a lesion. Alternatively, the contaminated area may be bypassed altogether. Examples of such approaches are transtracheal puncture with aspiration of lower respiratory secretions or suprapubic bladder puncture with aspiration of urine.

It is often impossible to collect an uncontaminated specimen, and decontamination procedures, cultures on selective media, or quantitative cultures must be used. Specimens collected by invasive techniques, particularly those obtained intraoperatively, require special attention. Enough tissue must be obtained for both histopathologic and microbiologic examination. Histopathologic examination is used to distinguish neoplastic from inflammatory lesions and acute from chronic inflammations. The type of inflammation present can guide the type of microbiologic examination performed. If, for example, a caseous granuloma is observed histopathologically, microbiologic examination should include cultures for mycobacteria and fungi. The surgeon should obtain several samples for examination from a single large lesion or from each of several smaller lesions. If an abscess is found, the surgeon should collect several milliliters of pus, as well as a portion of the wall of the abscess, for microbiologic examination. Swabs should be kept out of the operating room.

If possible, specimens should be collected before the administration of antibiotics. Above all, close communication between the clinician and the microbiologist is essential to ensure that appropriate specimens are selected and collected and that they are appropriately examined.

Microbiologic Examination

Direct Examination

Direct examination of specimens frequently provides the most rapid indication of microbial infection. A variety of microscopic, immunologic, and hybridization techniques have been developed for rapid diagnosis.

Sensitivity and Specificity

The sensitivity of a technique usually depends on the number of microorganisms in the specimen. Its specificity depends on how morphologically unique a specific microorganism appears microscopically or how specific the antibody or genetic probe is for that genus or species. For example, the sensitivity of Cram stains is such that the observation of two bacteria per oil immersion field (X 1,000) of a Gram-stained smear of uncentrifuged urine is equivalent to the presence of > 10^5 CFU/ml of urine. The sensitivity of the Gram-stained smear for detecting Gram-negative coccobacilli in cerebrospinal fluid from children with *Haemophilus influenzae* meningitis is approximately 75 percent because in some patients the number of colony-forming units per milliliter of cerebrospinal fluid is less than 10^4. At least 10^4 CFU of tubercle bacilli per milliliter of sputum must be present to be detected by an acid-fast smear of decontaminated and concentrated sputum.

An increase in the sensitivity of a test is often accompanied by a decrease in specificity. For example, examination of a Gram-stained smear of sputum from a patient with pneumococcal pneumonia is highly sensitive but also highly nonspecific if the criterion for defining a positive test is the presence of any Gram-positive cocci. If, however, a positive test is defined as the presence of a preponderance of Gram-positive, lancet-shaped diplococci, the test becomes highly specific but has a sensitivity of only about 50 percent. Similar problems related to the number of microorganisms present affect the sensitivity of immunoassays and genetic probes for bacteria, chlamydiae, fungi and viruses. In some instances, the sensitivity of direct examination tests can be improved by collecting a better specimen.

For example, the sensitivity of fluorescent antibody stain for Chlamydia trachomatis is higher when endocervical cells are obtained with a cytobrush than with a swab. The sensitivity may also be affected by the stage of the disease at which the specimen is collected. For example, the detection of herpes simplex virus by immunofluorescence, immunoassay, or culture is highest when specimens from lesions in the vesicular stage of infection are examined. Finally, sensitivity may be improved through the use of an enrichment or enhancement step in which microbial or genetic replication occurs to the point at which a detection method can be applied.

Techniques

For microscopic examination it is sufficient to have a compound binocular microscope equipped with low-power (1OX), high-power (40X), and oil immersion (1OOX) achromatic objectives, 10X wide-field oculars, a mechanical stage, a substage condenser, and a good light source. For examination of wet-mount preparations, a darkfield condenser or condenser and objectives for phase contrast increases image contrast. An exciter barrier filter, darkfield condenser, and ultraviolet light source are required for fluorescence microscopy.

For immunologic detection of microbial antigens, latex particle agglutination, coagglutination, and enzyme-linked immunosorbent assay (ELISA) are the most frequently used techniques in the clinical laboratory. Antibody to a specific antigen is bound to latex particles or to a heat-killed and treated protein A-rich strain of Staphylococcus aureus to produce agglutination. There are several approaches to ELISA; the one most frequently used for the detection of microbial antigens uses an antigen-specific antibody that is fixed to a solid phase, which may be a latex or metal bead or the inside surface of a well in a plastic tray. The test is then completed by adding a second antigen-specific antibody bound to an enzyme that can react with a substrate to produce a colored product.

Genetic probes are based on the detection of unique nucleotide sequences with the DNA or RNA of a microorganism. Once such a unique nucleotide sequence, which may represent a portion of a virulence gene or of chromosomal DNA, is found, it is isolated and inserted into a cloning vector (plasmid), which is then transformed into Escherichia coli to produce multiple copies of the probe. The sequence is then reisolated from plasmids and labeled with an isotope or substrate for diagnostic use. Hybridization of the sequence with a complementary sequence of DNA or RNA follows cleavage of the double-stranded DNA of the microorganism in the specimen.

The use of molecular technology in the diagnoses of infectious diseases has been further enhanced by the introduction of gene amplication techniques, such as the polymerase chain reaction (PCR) in which DNA polymerase is able to copy a strand of DNA by elongating complementary strands of DNA that have been initiated from a pair of closely spaced oligonucleotide primers. This approach has had major applications in the detection of infections due to microorganisms that are difficult to culture (e.g. the human immunodeficiency virus) or that have not as yet been successfully cultured (e.g. the Whipple's disease bacillus).

Culture

In many instances, the cause of an infection is confirmed by isolating and culturing microorganism either in artificial media or in a living host. Bacteria (including mycobacteria and mycoplasmas) and fungi are cultured in either liquid (broth) or on solid (agar) artificial media. Liquid media provide greater sensitivity for the isolation of small numbers of microorganisms; however, identification of mixed cultures growing in liquid media requires subculture onto solid media so that isolated colonies can be processed separately for identification. Growth in liquid media also cannot ordinarily be quantitated. Solid media, although somewhat less sensitive than liquid media, provide isolated colonies that can be quantified if necessary and identified. Some genera and species can be recognized on the basis of their colony morphologies.

In some instances one can take advantage of differential carbohydrate fermentation capabilities of microorganisms by incorporating one or more carbohydrates in the medium along with a suitable pH indicator. Such media are called differential media (e.g., eosin methylene blue or MacConkey agar) and are commonly used to isolate enteric bacilli. Different genera of the Enterobacteriaceae can then be presumptively identified by the colour as well as the morphology of colonies.

Culture media can also be made selective by incorporating compounds such as antimicrobial agents that inhibit the indigenous flora while permitting growth of specific microorganisms resistant to these inhibitors. One such example is Thayer-Martin medium, which is used to isolate *Neisseria gonorrhoeae.*

This medium contains vancomycin to inhibit Gram-positive bacteria, colistin to inhibit most Gram-negative bacilli, trimethoprim-sulfamethoxazole to inhibit *Proteus* species and other species that are not inhibited by colistin and anisomycin to inhibit fungi. The pathogenic *Neisseria* species, *N gonorrhoeae* and *N meningitidis*, are ordinarily resistant to the concentrations of these antimicrobial agents in the medium.

The number of bacteria in specimens may be used to define the presence of infection. For example, there may be small numbers ($< 10^3$ CFU/ml) of bacteria in clean-catch, midstream urine specimens from normal, healthy women; with a few exceptions, these represent bacteria that are indigenous to the urethra and periurethral region. Infection of the bladder (cystitis) or kidney (pyelone-phritis) is usually accompanied by bacteriuria of about $> 10^4$ CFU/ml. For this reason, quantitative cultures of urine must always be performed.

For most other specimens a semiquantitative streak method over the agar surface is sufficient. For quantitative cultures, a specific volume of specimen is spread over the agar surface and the number of colonies per milliliter is estimated. For semiquantitative cultures, an unquantitated amount of specimen is applied to the agar and diluted by being streaked out from the inoculation site with a sterile bacteriologic loop. The amount of growth on the agar is then reported semiquantitatively as many, moderate, or few (or 3+, 2+, or 1+), depending on how far out from the inoculum site colonies appear. An organism that grows in all streaked areas would be reported as 3+.

Chlamydiae and viruses are cultured in cell culture systems, but virus isolation occasionally requires inoculation into animals, such as suckling mice, rabbits, guinea pigs, hamsters, or primates. Rickettsiae may be isolated with some difficulty and at some hazard to laboratory workers in animals or embryonated eggs. For this reason, rickettsial infection is usually diagnosed serologically. Some viruses, such as the hepatitis viruses, cannot be isolated in cell culture systems, so that diagnosis of hepatitis virus infection is based on the detection of hepatitis virus antigens or antibodies. Cultures are generally incubated at 35 to 37°C in an atmosphere consisting of air, air supplemented with carbon dioxide (3 to 10 percent), reduced oxygen (microaerophilic conditions), or no oxygen (anaerobic conditions), depending upon requirements of the microorganism. Since clinical specimens from bacterial infections often contain aerobic, facultative anaerobic, and anaerobic bacteria, such specimens are usually inoculated into a variety of general purpose, differential, and selective media, which are then incubated under aerobic and anaerobic conditions.

The duration of incubation of cultures also varies with the growth characteristics of the microorganism. Most aerobic and anaerobic bacteria will grow overnight, whereas some mycobacteria require as many as 6 to 8 weeks.

Microbial Identification

Microbial growth in cultures is demonstrated by the appearance of turbidity, gas formation, or discrete colonies in broth; colonies on agar; cytopathic effects or inclusions in cell cultures; or detection of genus- or species-specific antigens or nucleotide sequences in the specimen, culture medium, or cell culture system.

Identification of bacteria (including mycobacteria) is based on growth characteristics (such as the time required for growth to appear or the atmosphere in which growth occurs), colony and microscopic morphology, and biochemical, physiologic, and, in some instances, antigenic or nucleotide sequence characteristics. The selection and number of tests for bacterial identification depend upon the category of bacteria present (aerobic versus anaerobic, Gram-positive versus Gram-negative, cocci versus bacilli) and the expertise of the microbiologist examining the culture. Gram-positive cocci that grow in air with or without added CO_2 may be identified by a relatively small number of tests.

The identification of most Gram-negative bacilli is far more complex and often requires panels of 20 tests for determining biochemical and physiologic characteristics. The identification of filamentous fungi is based almost entirely on growth characteristics and colony and microscopic morphology. Identification of viruses is usually based on characteristic cytopathic effects in different cell cultures or on the detection of virus- or species-specific antigens or nucleotide sequences.

Interpretation of Culture Results

Some microorganisms, such as *Shigella dysenteriae, Mycobacterium tuberculosis, Coccidioides immitis*, and influenza virus, are always considered clinically significant. Others that ordinarily are harmless components of the indigenous flora of the skin and mucous membranes or that are common in the environment may or may not be clinically significant, depending on the specimen source from which they are isolated. For example, coagulase-negative staphylococci are normal inhabitants of the skin, gastrointestinal tract, vagina, urethra, and the upper respiratory tract (i.e., of the nares, oral cavity, and pharynx). Therefore, their isolation from superficial ulcers, wounds, and sputum cannot usually be interpreted as clinically significant.

They do, however, commonly cause infections associated with intravascular devices and implanted prosthetic materials. However, because intravascular devices penetrate the skin and since cultures of an implanted prosthetic device can be made only after incision, the role of coagulase-negative staphylococci in causing infection can usually be surmised only when the microorganism is isolated in large numbers from the surface of an intravascular device, from each of several sites surrounding an implanted prosthetic device, or, in the case of prosthetic valve endocarditis, from several separately collected blood samples.

Another example, *Aspergillus fumigatus,* is widely distributed in nature, the hospital environment, and upper respiratory tract of healthy people but may cause fatal pulmonary infections in leukemia patients or in those who have undergone bone marrow transplantation. The isolation of *A fumigatus* from respiratory secretions is a nonspecific finding, and a definitive diagnosis of invasive aspergillosis requires histologic evidence of tissue invasion.

Physicians must also consider that the composition of microbial species on the skin and mucous membranes may be altered by disease, administration of antibiotics, endotracheal or gastric intubation, and the hospital environment. For example, potentially pathogenic bacteria can often be cultured from the pharynx of seriously ill, debilitated patients in the intensive care unit, but may not cause infection.

Serodiagnosis

Infection may be diagnosed by an antibody response to the infecting microorganism. This approach is especially useful when the suspected microbial agent either cannot be isolated in culture by any known method or can be isolated in culture only with great difficulty. The diagnosis of hepatitis virus and Epstein-Barr virus infections can be made only serologically, since neither can be isolated in any known cell culture system. Although human immunodeficiency virus type 1

(HIV-1) can be isolated in cell cultures, the technique is demanding and requires special containment facilities. HIV-1 infection is usually diagnosed by detection of antibodies to the virus.

The disadvantage of serology as a diagnostic tool is that there is usually a lag between the onset of infection and the development of antibodies to the infecting microorganism. Although IgM antibodies may appear relatively rapidly, it is usually necessary to obtain acute- and convalescent-phase serum samples to look for a rising titer of IgG antibodies to the suspected pathogen.

In some instances the presence of a high antibody titer when the patient is initially seen is diagnostic; often, however, the high titer may reflect a past infection, and the current infection may have an entirely different cause. Another limitation on the use of serology as a diagnostic tool is that immunosuppressed patients may be unable to mount an antibody response.

Antimicrobial Susceptibility

The responsibility of the microbiology laboratory includes not only microbial detection and isolation but also the determination of microbial susceptibility to antimicrobial agents. Many bacteria, in particular, have unpredictable susceptibilities to antimicrobial agents, and their susceptibilities can be measured in vitro to help guide the selection of the most appropriate antimicrobial agent.

Antimicrobial susceptibility tests are performed by either disk diffusion or a dilution method. In the former, a standardized suspension of a particular microorganism is inoculated onto an agar surface to which paper disks containing various antimicrobial agents are applied. Following overnight incubation, any zone diameters of inhibition about the disks are measured and the results are reported as indicating susceptibility or resistance of the microorganism to each antimicrobial agent tested. An alternative method is to dilute on a log2 scale each antimicrobial agent in broth to provide a range of concentrations and to inoculate each tube or, if a microplate is used, each well containing the antimicrobial agent in broth with a standardized suspension of the microorganism to be tested. The lowest concentration of antimicrobial agent that inhibits the growth of the microorganism is the minimal inhibitory concentration (MIC). The MIC and the zone diameter of inhibition are inversely correlated. In other words, the more susceptible the microorganism is to the antimicrobial agent, the lower the MIC and the larger the zone of inhibition. Conversely, the more resistant the microorganism, the higher the MIC and the smaller the zone of inhibition. The term susceptible means that the microorganism is inhibited by a concentration of antimicrobial agent that can be attained in blood with the normally recommended dose of the antimicrobial agent and implies that an infection caused by this microorganism may be appropriately treated with the antimicrobial agent.

The term resistant indicates that the microorganism is resistant to concentrations of the antimicrobial agent that can be attained with normal doses and implies that an infection caused by this microorganism could not be successfully treated with this antimicrobial agent. (A) Disk diffusion method. (B) Minimum inhibitory concentration (MIC) method. In the example shown, two different microorganisms are tested by both methods against the same antibiotic. The MIC of the antibiotic for the susceptible microorganism is 8 μg/ml. The corresponding disk diffusion test shows a zone of inhibition surrounding the disk. In the second sample, a resistant microorganism is not inhibited by the highest antibiotic concentration tested (MIC > 16 μg/ml) and there is no zone of inhibition surrounding the disk. The diameter of the zone of inhibition is inversely related to the MIC.

Antimicrobial Chemotherapy

Classification

Antimicrobials can be classified by at least three different schemes:

1. Effects on cells
2. Range of activity
3. Sites of activity

This page will examine these different classification schemes and describe several examples of each type of antimicrobial. The mechanisms by which organisms become resistant to these agents will also be discussed. Finally, toxicologic properties of antimicrobial chemotherapy will be described.

Antimicrobial Effects on Cells

Antimicrobials can be divided into two classifications based upon their effects on target cells. Drugs that actually kill microorganisms are termed *bactericidal*. Drugs that only inhibit the growth of microorganisms are termed *bacteriostatic*. The decision to use a bactericidal or bacteriostatic drug to treat infection depends entirely upon the type of infection. For example, bactericidal drugs will only kill cells that are actively growing. Bacteriostatic drugs, in comparison, will only inhibit the growth of cells; ultimate elimination of the organisms is dependent upon host phagocytic activity. Some examples of bactericidal and bacteriostatic drugs are listed below.

Range of Activity

Antimicrobials can also be classified by their range of activity. In general, five classifications can be described. The first of these is termed *narrow spectrum*. Narrow spectrum drugs, as the name implies, are only active against a relatively small number of organisms. In general, narrow spectrum antibiotics are effective against Gram-positive organisms. The second classification is termed *moderate spectrum*. These drugs are generally effective against the Gram-positives and most systemic, enteric and urinary tract Gram-negative pathogens. The beta-lactam antibiotics (penicillin, ampicillin, cephalosporins, etc.) belong in a third classification, *narrow and moderate spectrum* because some members are only effective against Gram-positive organisms while other members can also kill certain Gram-negative bacteria. A fourth classification is termed *broad spectrum*. These drugs are effective against all prokaryotes with two exceptions: *Mycobacteria* and *Pseudomonas*. The fifth group includes those drugs that are effective against *Mycobacteria*.

Antimicrobial Chemotherapy for Legionnaires Disease

Paul H. Edelstein, MD + Author Affiliations

University of Pennsylvania Medical Centre; Philadelphia, PA 19104-4283 Note: Dr. Edelstein has received funding for research and related academic purposes from manufacturers of both macrolide and quinolone antibiotics. Acknowledgment: The author thanks Richard D. Meyer for reviewing the manuscript and suggesting changes. Requests for Reprints: Paul H. Edelstein, MD, Clinical Microbiology Laboratory, 4 Gates, University of Pennsylvania Medical Centre, 3400 Spruce Street, Philadelphia, PA 19104-4283; e-mail, phe@mail.med.upenn.edu. Erythromycin became the de facto standard of treatment for Legionnaires disease soon after the first recognized epidemic of this disease occurred 22 years ago. However, it is now time for a change in therapy because of increasing evidence of the superiority of newer antimicrobial agents. The conclusion that erythromycin is the drug of choice for Legionnaires disease was based on unavoidably imperfect retrospective analysis of treatment outcomes from the epidemic, which showed the lowest overall fatality rates with either erythromycin or tetracycline therapy.

No prospective comparative studies with adequate statistical power of the treatment of Legionnaires disease have been published. Because of the relative rarity of Legionnaires disease and its widely varying outcomes, it is unlikely that an adequate comparative clinical trial of the disease will ever be completed; such a study would need to have 300 to 900 patients with the disease in each experimental group to have adequate statistical power. This means that decisions about new therapy must be based on laboratory experiments and limited clinical data.

Case-fatality rates for treated Legionnaires disease depend on various factors, but the chief determinants seem to be 1) the premorbid immune and general health status of the patient and 2) whether the disease is community-acquired or nosocomial. Higher fatality rates for nosocomial infection are probably due to host-related rather than organism-related factors. Previously healthy patients with community-acquired Legionnaires disease who do not require hospitalization respond well to erythromycin therapy about 95% to 99% of the time.

Quite a different picture emerges for immunocompromised patients with Legionnaires disease or for patients with nosocomial Legionnaires disease. Fatality rates of up to 50% have been reported for treated patients in these categories, with a very wide range; fatality rates as high as 80% have been found in untreated patients. Prolonged illness before therapy and respiratory failure caused by pneumonia are especially poor prognostic factors.

Relapse after cessation of erythromycin therapy can be a serious problem in some immunosuppressed patients and can require prolonged and repeated courses of therapy. The occasional ineffectiveness of erythromycin therapy was recently highlighted in a report of fatal *Legionella micdadei* pneumonia in an immunocompromised patient who died despite receiving multiple courses of erythromycin therapy.

In addition to giving rise to potentially poor outcomes in patients with severe Legionnaires disease, erythromycin causes numerous unpleasant side effects. These include high rates of phlebitis and infusion pain with intravenous administration and frequent gastrointestinal problems with oral administration. Large intravenous infusion volumes are often required to increase patient tolerance, and this can exacerbate fluid management problems in critically ill patients. Other possible problems include cardiac arrhythmias and conduction abnormalities, hepatitis, hearing loss, interference with the metabolism of numerous drugs, and poor patient tolerance overall. Of adults who receive outpatient therapy with erythromycin, 12% to 33% do not complete therapy because of intolerance; this contrasts with the much lower dropout rates for most other antibiotics.

Because of adverse effects, hospitalized patients receiving intravenous erythromycin may have significantly increased lengths of stay and increased costs of hospitalization compared with patients receiving other agents. In some outpatients, the frequency of adverse effects may mean that the long courses of therapy required to cure Legionnaires disease are not completed, resulting in relapse.

If valid comparative clinical studies cannot be done, how can the value of new drugs be assessed? Luckily, certain types of laboratory studies have correlated very well with clinical response to infection. For example, animal models and intracellular infection models correctly predict that erythromycin is clinically effective and that cefoxitin is not. The success of these correlations is based on the intracellular location of *L. pneumophila* and the ability of drugs to achieve high, bioactive, intracellular concentrations.

It is less certain that an antimicrobial agent that is highly active in such experiments is more effective clinically than a drug (such as erythromycin) that is marginally effective in laboratory tests. The lack of controlled clinical trials means that this question is not easily answered. One retrospective study showed that pefloxacin is more active clinically than erythromycin; this finding correlates with the superior laboratory activity of pefloxacin.

In intracellular models of *L. pneumophila* infection, bacterial growth is inhibited by erythromycin and rifampin but recurs when the drugs are removed from the cells. Thus, these drugs differ from many newer fluoroquinolones, which can kill most, if not all, of the bacteria present in the infected cells and can be removed from the cells without causing rapid bacterial regrowth.

Against intracellular *L. pneumophila*, levofloxacin, sparfloxacin, and trovafloxacin are the most active fluoroquinolone antimicrobial agents in the laboratory that are licensed by the U.S. Food and Drug Administration (FDA). Azithromycin is also more active than erythromycin against intracellular *L. pneumophila*.

Studies of animal models of *L. pneumophila* pneumonia confirm the superior activity of most fluoroquinolone antimicrobial agents and azithromycin compared with erythromycin. In these experiments, a single low dose of azithromycin not only is curative but also markedly reduces lung concentrations of *L. pneumophila*. In contrast, multiple, high doses of erythromycin are needed to achieve a similar cure rate without the benefit of rapid lung sterilization.

Extremely low dosages of trovafloxacin or sparfloxacin also have had curative and sterilizing effects in animal studies. The failure of erythromycin treatment to completely eradicate *L. pneumophila* from guinea pigs with experimental pneumonia correlates with disease relapse in some heavily immunosuppressed patients after cessation of erythromycin therapy.

Uncontrolled clinical trials show that many new fluoroquinolone antimicrobial agents effectively treat both community-acquired and nosocomial Legionnaires disease; these agents include ciprofloxacin, pefloxacin, ofloxacin, levofloxacin, and trovafloxacin (trovafloxacin has been studied against community-acquired disease only). Similar clinical data exist for azithromycin.

On the other hand, treatment failures have been reported for these newer antimicrobial agents, as they have for erythromycin, and use of these agents is not a guarantee of therapeutic success. Laboratory evidence suggests that the most active fluoroquinolone antimicrobial agents are trovafloxacin, sparfloxacin, and pefloxacin. Levofloxacin, ofloxacin, and probably grepafloxacin possess intermediate but still high activity.

Ciprofloxacin is slightly less active than the other compounds but is still more active than erythromycin. Azithromycin is substantially more active in animal experiments than is clarithromycin. Azithromycin (intravenous formulation), levofloxacin, and trovafloxacin are currently licensed by the FDA for the treatment of Legionnaires disease, as are erythromycin and dirithromycin.

In my opinion, azithromycin or one of the more active fluoroquinolones should be used in preference to erythromycin for the treatment of Legionnaires disease in hospitalized and immunocompromised patients. The potentially high mortality and morbidity rates of the disease in such patients, combined with the experimental evidence discussed above, is reason enough to use more active antimicrobial agents in this setting. In addition to producing a potentially better outcome, these agents will often improve patient convenience as a result of fewer side effects and shorter durations of therapy. Adding rifampin to the above therapies probably provides little further benefit; no convincing laboratory data show that adding rifampin to fluoroquinolone or more active macrolide therapy improves bacterial killing. In our institution, levofloxacin (500 mg/d for 7 to 10 days) or azithromycin (500 mg/d for 3 days; 500 mg on day 1 and 250 mg/d for 4 more days; or 500 mg/d for 5 days) are recommended for the treatment of inpatients with Legionnaires disease.

For nonimmunocompromised outpatients with community-acquired Legionnaires disease, erythromycin or tetracycline therapy remain perfectly acceptable. These drugs are much less expensive than the newer agents, and treatment outcomes in this population are likely to be the same with new and old therapies. Ten to 14 days of treatment with erythromycin or doxycycline should be sufficient for such patients.

Drug intolerance or expected poor compliance with a longer duration of therapy are the main reasons to prescribe the newer agents. As with the newer antimicrobial agents, the possibility that a patient whose illness seems unresponsive to these other agents has pneumonia caused by an agent other than *Legionella* species, especially an antimicrobial-resistant bacterium, should be considered.

10

Pest Control Tactics

The history of pest control probably began with the first human who ever swatted a mosquito or picked off a louse. From the fossil record, we know that all major taxa of biting flies and external parasites already existed by the time *Homo sapiens* first appeared on earth. *Phthirus* and *Pediculus*, the two genera of lice that feed on humans, have a host range that is limited to primates (apes and monkeys). And we suspect that human fleas (*Pulex irritans*) and bed bugs (*Cimex lectularius*) adopted cave-dwellers as hosts because these insects are most closely related to other species that live on bats.

But since our primitive ancestors were hunters and gatherers, they probably found that insects were more useful as food than they were troublesome as pests. (Even today, people in some primitive cultures eat the lice they pick from one another's hair). It was probably not until the dawn of organized agriculture, when insects attacked the plants we grew for food, that we first recognized them as a potential threat to our own survival.

Pest control tactics were mentioned occasionally in writings of the ancient Chinese, Sumerian, and Egyptian scholars. Many of these tactics were embedded in religion or superstition, but a few had real scientific merit.

Predatory ants, for example, were used in China as early as 1200 B.C. to protect citrus groves from caterpillars and wood boring beetles. Ropes or bamboo sticks tied between adjacent branches helped the ants move easily from place to place. A passage in Homer's Iliad (8th century B.C.) describes the use of fire to drive locusts into the sea, and the ancient Egyptians organized long lines of human drovers to repel swarms of invading locusts. Pythagorus, a Greek philosopher and mathematician, was credited with clearing malaria from a Sicilian town during the 6th century B.C. by instructing its residents to drain the marshes. Chemical substances that purportedly killed or repelled insects were in common usage. Many of these were of questionable value, but some worked, and a few are still in use today. Some of the inorganic compounds, such as sulfur and arsenic, have well-established insecticidal activity. And modern science has only recently come to recognize that many plant extracts used by ancient apothecaries (e.g, lemon oil, wormwood, hellebore, fleabane, etc.) do indeed contain compounds with useful activity against insects.

There was very little progress in pest control during the dark ages. Ignorance and superstition abounded. For what it was worth, St. Bernhard excommunicated the flies of his parish in 1121. In a book entitled "Natural History", Ferrante Imperato (1599) gave a prescription for eliminating flies from a dwelling:

> *"... draw the image of a fly... on a copper plate during the second half of the constellation of Pisces... then bury it in the centre of your house (during) the first half of the constellation of Taurus."*

With the Renaissance, people began to view insects less as a punishment from God and more as members of a natural world that could be studied and controlled. A renewed interest developed in insects both as organisms and as pests. More accurate observations of natural history and behaviour led to more inventive control practices. Hand labour was used extensively in early pest control, but cultural, physical, and chemical practices also evolved. Franz Ernst Brückmann, a German physician who lived in the early 1700's, designed the first mechanical traps for various insects. His fly traps consisted of a wooden box baited with a sweet attractant and equipped with a spring-loaded lid. When several flies had settled inside the trap, the lid was snapped shut to trap the flies inside. The volume of the trap was then decreased by sliding the ends of the box together and squashing the flies against their bait. Brückmann also designed flea traps. They were hollow, perforated cylinders, baited with blood or honey, and worn around the neck as a pendant. For a time, these flea traps, crafted from ivory or silver and often ornate in design, were popular fashion accessories worn by the aristocracy of western Europe.

Since the late 1800's, entomologists and chemists have made outstanding progress in the technology of pest control. Today's arsenal of weapons is large and diverse, encompassing legal, cultural, physical, genetic, and biological tactics, in addition to the well-known chemical pesticides. In general, all of these tactics work in at least one of the following ways:

- They kill the pest directly — usually by exposing it to lethal substances or unsuitable environmental conditions.
- They reduce the reproductive potential of a pest population — often by modifying its environment (biotic or abiotic) or by restricting its movement.
- They modify the pest's behaviour to make it less troublesome (attract it, repel it, confuse it, exclude it, mislead it).

Most of the control tactics that are commonly used today can be grouped into two broad categories: natural controls and artificial controls. By definition, a natural control may be any environmental factor that keeps a pest population below its economic injury level. Examples might include geographic barriers, cold temperatures, or natural enemies that keep population growth in check. Artificial controls, on the other hand, employ products or processes of human origin to modify a pest's distribution, behaviour or physiology. Fly swatters and insecticides are two obvious examples. But as is often true in biology, there are some control strategies that seem to straddle the line, exhibiting characteristics of both groups or not fitting neatly into either category. In fact, some of our most effective tactics are natural controls that we can improve, enhance, accelerate, or augment by appropriate human intervention.

Cultural Control

Surprisingly simple modifications of a pest's environment or habitat often prove to be effective methods of pest control. As a group, these tactics are usually known as cultural control practices because they frequently involve variations of standard horticultural, silvicultural, or animal husbandry practices.

Since these control tactics usually modify the relationships between a pest population and its natural environment, they are also known, less commonly, as ecological control methods. Simplicity and low cost are the primary advantages of cultural control tactics, and disadvantages are few as long as these tactics are compatible with a farmer's other management objectives (high yields, mechanization, etc.). Unfortunately, there are still a wide variety of insect pests that cannot be suppressed by cultural methods alone.

Crop rotation is one of the oldest and most effective cultural control strategies. Growing a single crop year after year in the same field gives pest populations sufficient time to become established and build up to damaging levels. Rotating the field to a different type of crop can break this cycle by starving pests that cannot adapt to a different host plant. Farmers in the Midwest, for example, can reduce populations of wireworms (Elateridae) and rootworms (*Diabrotica* spp.) in corn fields by switching to oats, wheat, or legumes. Similarly, the clover root curculio (*Sitona hispidula*) that feeds exclusively on legumes can be eliminated by switching from clover or alfalfa to corn or small grains. Rotation schemes have also proven successful for controlling pests in pasture lands. Texas cattle fever, a protozoan disease transmitted by cattle ticks, *Boophilus annulatus*, has been eliminated from the southern United States due, in part, to a diligent campaign of pasture rotation designed to protect beef and dairy herds from tick infestation.

Crop rotation schemes work because they increase the diversity of a pest's environment and create discontinuity in its food supply. As a rule, rotations are most likely to be practical and effective when they are used against pests that:

a. attack annual or biennial crops
b. have a relatively narrow host range
c. cannot move easily from one field to another, and
d. are present before the crop is planted.

Intercropping (also known as mixed cropping) is another way to reduce pest populations by increasing environmental diversity. In some cases, intercropping lowers the overall attractiveness of the environment, as when host and non-host plants are mixed together in a single planting. But in other cases, intercropping may concentrate the pest in a smaller, more manageable area so it can be controlled by some other tactic. Strips of alfalfa, for example, are sometimes interplanted with cotton as a trap crop for lygus bugs (Miridae). The alfalfa, which attracts lygus bugs more strongly than cotton, is usually treated with an insecticide to kill the bugs before they move into adjacent fields of cotton.

In some crops, it is possible to create discontinuity in the pest's food supply simply by altering the time of year for planting or harvesting. This strategy, often known as phenological asynchrony, allows farmers to manage their crop so it remains "out of phase" with pest populations. Sweet corn, for example, can escape most injury from corn earworms (*Helicoverpa zea*) if it is planted in early spring and harvested before larvae mature. In the Midwest, farmers delay planting winter wheat until after the "fly-free date" to protect their crop from injury by the Hessian fly (*Mayetiola destructor*). These planting dates are calculated to ensure that wheat is not mature enough to attract egg-laying flies, yet still has enough warm weather to grow sufficiently before winter. Careful timing of harvest dates in alfalfa can cause high mortality in populations of alfalfa weevil (*Hypera postica*) and alfalfa caterpillar (*Colias eurytheme*) by removing edible foliage before the larvae have completed development. Harvest timing is often the most practical method available to foresters for controlling bark beetle (Scolytidae) infestations in pine plantations. Since trees become more susceptible to beetle outbreaks as they grow older, good management dictates that timber stands be harvested for lumber before the trees reach full maturity.

Managed application of water or fertilizer can have a big impact on the survival of pest populations in some crops. Annual flooding, for example, is a cultural practice that eliminates many potential pests in cranberry bogs. Regular rainfall (or overhead irrigation) can significantly reduce infestations of twospotted spider mites (*Tetranychus urticae*) in tree fruits. Good irrigation and appropriate fertilization keeps plants healthy, vigorous, and more resistant to insect injury. It is not unusual for small amounts of injury to actually stimulate compensatory growth in healthy plants.

Sanitation is another cultural control strategy that may be highly effective for some pests. Removing crop debris from cotton fields after harvest eliminates overwintering populations of pink bollworms (*Pectinophora gossypiella*), European corn borers (*Ostrinia nubilalis*), and sugarcane borers (*Diatraea saccharalis*). Collecting dropped fruit from beneath an apple tree reduces the next season's population of apple maggots (*Rhagoletis pomonella*), codling moths (*Cydia pomonella*), and plum curculio (*Conotrachelus nenuphar*). Shredding or burning the pruning wood from a peach orchard kills shothole borers (*Scolytus rugulosus*) and lesser peachtree borers (*Synanthedon pictipes*) that would otherwise emerge and reinfest the orchard. Clean cultivation is often recommended as a way to eliminate shelter and/or overwintering sites for pest populations. Simply tilling or plowing a corn field before winter may disrupt a pest's life cycle by causing mechanical injury, by increasing exposure to lethal cold temperatures, by intensifying predation by birds or small mammals, or by burying the pests deep beneath the soil surface. Populations of corn earworms and European corn borers have been greatly reduced in recent years by community-wide efforts to plow under corn stubble after harvest.

Although it works well in corn and cotton, clean cultivation is not always the best solution for eliminating pest populations. In some cases, ground cover or crop debris shelters natural enemies that are important members of the agroecosystem. Without adequate shelter these beneficial populations would die or move away.

Biological Control

Natural control strategies that employ biological agents for pest suppression are generally classified as biological control tactics. In conventional usage, this term usually refers to the practice of rearing and releasing natural enemies: parasites, predators, or pathogens. A slightly broader definition of "biocontrol" includes any related management activity that is designed to protect or conserve natural enemies.

Biocontrol agents include a wide variety of life forms, including vertebrates, invertebrates, fungi, and microorganisms. These beneficial species are common in most natural communities and, although their presence is often unnoticed, they help maintain the "balance of nature" by regulating the density of their host or prey population. Insect species often become "pests" when this ecological balance is disrupted by natural events or human intervention. Biological pest control strives to reestablish this balance in one of three ways:

Importation. Foreign exploration is conducted to identify and collect natural enemies in the country from which an exotic pest has been introduced. Following the discovery of a potential biocontrol agent, it undergoes extensive evaluation to insure that its ecology and host range are compatible with the community to which it will be introduced and that it will not become a pest once it is released. Suitable candidates are reared and released in the new habitat in hopes that they will become established and suppress the pest population.

1. *Conservation.* A variety of management activities can be used to optimize the survival and/or effectiveness of natural enemies. Conservation activities might include reducing or eliminating insecticide applications to avoid killing natural enemies, staggering harvest dates in adjacent fields or rows to insure a constant supply of hosts (prey), or providing shelter, over-wintering sites, or alternative food sources to improve survival of beneficial species.

2. *Augmentation.* Natural enemies that are unable to survive and/or persist in a new environment can sometimes be reared in large numbers and periodically released to suppress a pest population. In some cases, small numbers of a beneficial species are released in several critical locations to suppress local pest outbreaks (an inoculative release). In other cases, larger numbers are released in a single location to flood the pest population with natural enemies (an inundative release).

Examples of Biocontrol

Just because an insect is a predator or a parasite does not necessarily make it a good biological control agent. Since the early 1900's over 600 species of beneficial insects have been introduced into the United States. Of these, only about 20% have been outright successes as biological control agents; another 35% have been partially successful, and the other 45% never became established or failed to have any significant effect on pest populations. From these successes and failures, entomologists have concluded that introduced biocontrol agents are most effective when they exhibit the following characteristics:

Narrow host range. Generalized predators may be good natural enemies but they don't kill enough pests when other types of prey are also available.

1. *Climatic adaptability.* Natural enemies must be able to survive the extremes of temperature and humidity that they will encounter in the new habitat.
2. *Synchrony with host (prey) life cycle.* The predator or parasite should be present when the pest first emerges or appears.
3. *High reproductive potential.* Good biocontrol agents produce large numbers of offspring. Ideally, a parasite completes more than one generation during each generation of the pest.
4. *Efficient search ability.* In order to survive, effective natural enemies must be able to locate their host or prey even when it is scarce. In general, better search ability results in lower pest population densities.
5. *Short handling time.* Natural enemies that consume prey rapidly or lay eggs quickly have more time to locate and attack other members of the pest population. Small populations of efficient natural enemies may be more effective biocontrol agents than larger populations of less efficient species.

6. *Survival at low host (prey) density.* If a natural enemy is too efficient, it may eliminate its own food supply and then starve to death. The most effective biocontrol agents reduce a pest population below its economic threshold and then maintain it at this lower equilibrium level.

Just as predators and parasites may serve as effective biocontrol agents for insect pest populations, so insect herbivores have also proven effective as agents for controlling weed populations. Feeding injury by herbivores may weaken a plant or increase its susceptibility to disease, desiccation, freezing temperatures, or other herbivores. In any case, an injured plant is likely to be a poor competitor, and in time, beneficial species will have an opportunity to reclaim the vacated space.

Insect Weed-eaters

The Klamath weed (*Hypericum perforatum*) and the prickly pear cactus (*Opuntia* spp.) are two well-known examples of successful weed control using insect herbivores. Biological control is a particularly appealing pest control alternative because, unlike most other tactics, it does not always have to be reapplied each time a pest outbreak occurs. Once natural enemies are released into a new environment, there is a good chance they will become established and provide a self-perpetuating form of control. Biological control is also the only control tactic that increases, rather than decreases, the species diversity within an agroecosystem. This increased diversity often results in greater stability because wild fluctuations in population density are less common in communities with a diverse food web.

Biological control is not a "quick fix" for most pest problems. Natural enemies usually take longer to suppress a pest population than other forms of pest control and farmers often regard this as a disadvantage. It also may be difficult to "integrate" natural enemies into a crop or commodity when pesticides are still in use. Beneficial insects are often highly sensitive to pesticides and their resurgence (recovery to pre-spray densities) is usually much slower than that of pest populations. Rapid pest resurgence often leads to a vicious cycle of continued chemical usage that prevents natural enemies from ever becoming reestablished.

Perhaps the greatest potential for future progress in biological control lies in improving the success of microbial pathogens. Many of these organisms are highly desirable as biocontrol agents: they attack a narrow range of insect hosts, they are not hazardous to humans or domestic animals, and they do not pose a threat to the environment.

As a group, though, pathogens have never proven to be quite as reliable as other forms of pest control. They are vulnerable to desiccation, ultraviolet radiation, and high temperatures. They may not survive long enough in the environment to encounter a host, and even if they survive, their virulence may be too low to overcome the host's defenses.

In the past few years, however, there has been encouraging progress in the development of microbial insecticides. These are commercial suspensions of spores, toxins, or virus particles that can be mixed with water and sprayed onto crops just like conventional insecticides. In many cases, microbial insecticides are better than conventional insecticides because they suppress pest populations without eliminating natural populations of predators and parasites.

A great deal of research is still needed before we can begin to benefit from the full potential of biocontrol. Fortunately, new developments in biotechnology may soon enable us to create new strains of microbial pathogens that are more virulent, easier to mass produce, and less sensitive to temperature and humidity. Work is currently underway to transform a common soil microbe, *Pseudomonas fluorescens*, into a pathogen of soil-dwelling insects by implanting the delta-endotoxin gene from *Bacillus thuringiensis*. Clearly, there are major changes in biocontrol that lie just over the horizon.

Breeding for Host Resistance

Breeding plants (or animals) for resistance to insects is really just another form of biological pest control. Rather than finding insects to attack the pests, breeders look for genetic traits (or combinations of traits) that reduce an organism's susceptibility to attack or injury by its insect pests. This idea was first tested in the 1870's by C. V. Riley, an entomologist who successfully fought a French outbreak of phylloxera (an aphid-like pest of grapes) with resistant North American rootstocks. Indeed, much of the pioneering work in this field has been done with crop plants because they are usually cheaper to breed and easier to manage than livestock. Recent innovations in genetic engineering, however, are likely to accelerate development of resistant genotypes in poultry, swine, and other commercial livestock.

In general, there are three approaches that plant breeders use to develop resistant cultivars:

1. *Antibiosis*. Plants produce a wide variety of defensive compounds (allelochemicals) that protect them from herbivores.

These compounds may reduce growth, inhibit reproduction, alter physiology, delay maturation, or induce various physical or behavioral abnormalities in herbivores. By purposely selecting for plants with high levels of allelochemicals, or by breeding such plants with less resistant ones, it is often possible to develop new cultivars that resist pest injury yet still retain desirable horticultural characteristics.

2. *Antixenosis.* A physical or chemical property of a plant can make it so unpalatable that it is largely protected from herbivore attack. This type of resistance is often known as nonpreference. It may involve the presence of feeding repellents (or the absence of feeding attractants), or it may involve physical traits such as hairs, waxes, or a thick, tough epidermis that do not provide the pest with a desirable feeding substrate. Alfalfa, for example, has been bred with hairy leaves to deter feeding by the spotted alfalfa aphid.
3. *Tolerance.* Some plant genotypes are simply able to “tolerate” injurious insects better than others. Tolerant cultivars may be exposed to the same pest populations as susceptible ones, but they do not suffer as much injury. Many varieties of field corn, for example, have fairly narrow, brittle stalks. When attacked by European corn borers, these stalks are further weakened and break easily in a windstorm. This wind damage, known as lodging, makes the corn hard to pick with mechanical harvesters. Plant breeders have largely resolved this problem by breeding for corn with thicker, stronger cornstalks. These tolerant cultivars are still attacked by corn borers, but they “stand up” to the injury and insure a harvestable crop.

Developing resistant cultivars by traditional breeding methods can be a slow and uncertain process. But new advances in biotechnology now make it possible to pluck genes from one organism and insert them into the cells of a completely different species. For example, the delta-endotoxin gene from *Bacillus thuringiensis*, an insect pathogen, has already been isolated and inserted into crop plants (notably cotton and tobacco). Cells of the genetically engineered plants can produce their own toxin, making them “resistant” to certain insect herbivores (e.g., Lepidoptera). Although gene-splicing techniques are still in their infancy, they offer exciting new possibilities for expanding the diversity of resistant plant genotypes.

Legal and Regulatory Control

At first, it may seem ludicrous to even contemplate the use of legal measures as pest control strategies — it's easy enough to pass laws, but how do you train the insects to obey them? In reality, legal sanctions are aimed, not at the insects themselves, but at a range of human behaviors that are most likely to affect the dynamics of pest populations. Legal control tactics, therefore, include all forms of legislation and regulation that might prevent establishment or reduce spread of an insect population.

The Plant Quarantine Act of 1912 was the first legal action taken in the United States to prevent the introduction of pests from foreign countries. This law, and others that followed it, established a network of inspection stations at major ports of entry and gave the federal government authority to organize border quarantines, to inspect all agricultural products, and to restrict entry of any infested goods. Today, these inspections stations operate under the jurisdiction of the Animal and Plant Health Inspection Service (APHIS), a branch of the U. S. Department of Agriculture. APHIS operates about 85 inspection facilities, employs 400-500 inspectors, and intercepts over 20,000 potential pest introductions in an average year. It also operates the Plant Quarantine Training Centre where inspectors from all over the world are trained to spot problems before they cross international borders. These agents, now representing nearly 40 countries, cooperate in efforts to limit the spread of indigenous pests under an international agreement that requires inspection and certification of nearly all agricultural commodities prior to export.

Federal import/export laws have been enacted to establish standards for a wide range of agricultural products. Most of the fruits and vegetables imported to the U.S., for example, must receive some form of fumigation, heat treatment, controlled atmosphere storage, or radiation exposure before they are released from quarantine. These post-harvest treatments are designed to kill whatever pest species may be present without affecting taste or quality of the produce.

Despite the best efforts of APHIS inspectors, some pests still slip into the U. S. and become established (e.g., cereal leaf beetle), or move naturally across international borders (e.g., imported fire ants and Africanized honey bees). When this happens, APHIS cooperates with state and local governments to establish domestic quarantines for limiting the movement of these pests within our borders. Whenever appropriate, eradication programs may be launched together with quarantine actions in an effort to destroy the entire pest population while it is still small. Inspections and quarantines are also operated by some state and local government agencies. The California Border Patrol, for example, is notorious for its interceptions of uncertified agricultural products at the state line. The rigid laws that prohibit private citizens from bringing a bag of oranges into southern California are generally given credit for excluding numerous pest species that might have caused millions of dollars in losses.

The price tag for these prevention efforts is only a fraction of the estimated $350 million Californians spent to clean up their most recent infestation of the Mediterranean fruit fly (*Ceratitis capitata*). Since 1915, nearly 300 pest species have been under some sort of domestic quarantine. The success of these programs has been erratic. In some cases (e.g., screwworm flies, khapra beetles, and parletoria date scale), eradication programs have managed to eliminate the pests. But for every noteworthy success, there has been a complete failure (witness the cereal leaf beetle, the gypsy moth, and the European corn borer). In general, eradication efforts are most successful when the pest population is relatively small, does not disperse rapidly, and is highly susceptible to control tactics.

Even if a newly established pest population cannot be eradicated, it can often be slowed down a bit by regulating the movement of its host plants or animals, prohibiting sales of infested products, and inspecting all shipments of affected goods. This type of containment strategy usually gives growers and producers a short reprieve before invasion of the pest and buys a little more time for researchers to find better methods of control. By using the containment approach in Texas during a 1971 outbreak of Venezuelan equine encephalitis, entomologists and veterinarians had enough time to reduce populations of mosquito vectors and immunize nearly 3 million horses before the disease could spread into other states.

Licensing and certification are important regulatory tools used in some commodities to ensure that infested or contaminated material is not sold commercially or used as breeding stock. Maintaining disease-free mother plants in strawberries, for example, can often prevent the spread of insect-borne disease during the growing season. Certified seed or plant material is produced under strict guidelines established by commodity organizations in their own interests. These products may be more expensive initially, but eventual savings in pest control usually provide a good return on the investment.

Eugenics and Birth Control

Some control tactics are designed to suppress a pest population by altering its genetic makeup and/or reducing its reproductive potential. As a group, these tactics are frequently known as genetic controls because they affect the accuracy or efficiency with which a pest species passes its genetic material (DNA) from one generation to the next. Genetic control usually works in one of two ways: either by causing (inducing) reproductive sterility, or by incorporating new and potentially deleterious genes (or alleles) into the genetic makeup of a pest population. In effect, some members of a pest species are transformed into biological time bombs that eventually destroy other members of their own species. Because of the self-destructive nature of these tactics, they are sometimes called autocidal control.

Insects can be sterilized by exposing them to certain chemical agents (chemosterilants) or to non-lethal levels of ionizing radiation (X-rays or gamma rays). Chemosterilants are really a form of chemical control. They usually work by blocking the onset of sexual maturity, by inhibiting the production of eggs and/or sperm, or by damaging the chromosomes. These compounds will be covered more fully in the section on semiochemical insecticides.

Exposure to radiation also damages chromosomes (usually by breakage or mutation). Since cells with damaged chromosomes cannot divide correctly, they do not form normal gametes or produce viable offspring. Although the susceptibility of each insect species is different, a proper dose of radiation administered at an appropriate stage of development (usually to pupae) can often induce sterility without causing other deleterious side effects. Sterile individuals, reared in large numbers and released into the environment, can mate with "normal" individuals but they produce no viable offspring. The more sterile individuals released, the fewer "normal" matings are likely to occur.

The effects of radiation on insect development was first studied in 1916, but this knowledge was not put to practical use until the 1930's when E. F. Knipling suggested that it might be possible to suppress pest populations by flooding the environment with large numbers of males that had been rendered infertile by irradiation. By maintaining a constant population of sterile males that was large in comparison to the number of virgin females, Knipling calculated that the number of "normal" matings would decrease each generation until the population was forced into extinction.

The first large-scale test of Knipling's sterile male technique was conducted in 1954 on the island of Curaçao against the screwworm, *Cochliomyia hominovorax*, a Dipteran pest of livestock. After just 13 weeks (four to five generations of the fly) and release of nearly a million sterile males, the screwworm population was completely eradicated from this tiny Caribbean island.

The dramatic success of Knipling's first project led to larger screwworm eradication programs in the southeastern and southwestern United States and to intense interest in the sterile male technique for eradication of other pest species, including Mediterranean fruit flies (*Ceratitis capitata*), boll weevils (*Anthonomus grandis*), horn flies (*Haematobia irritans*), tsetse flies (*Glossina* spp.), pink bollworms (*Pectinophora gossypiella*), and codling moths (*Cydia pomonella*).

Overall, the sterile male technique has had a spotty record of achievement. Despite isolated successes, no other eradication program has ever measured up to the outstanding performance of the screwworm project. From a theoretical standpoint, the use of sterile males is highly attractive because it is non-polluting, species specific, and becomes more effective as the target population decreases. But in practice, the technique simply doesn't work unless a pest population meets the following criteria:

1. *Easy to mass-produce.* A sterile insect release program cannot even be considered unless large numbers of insects can be produced at relatively low cost.
2. *Females mate only once.* This characteristic is uncommon among insects, but it is important for success of the sterile male technique because multiple matings only increase the probability that a female will receive sperm from a fertile male.
3. *Males can be sterilized without loss of competitive vigor.* Sterile males must be able to compete successfully for mates with normal (fertile) males. Sterile males must be physiclly and behaviorally identical to normal males.
4. *Low initial population.* The pest population must be small enough so the initial release of sterile males will outnumber the population of normal males. In some cases, alternative control tactics may be used to reduce initial populations to practical levels.
5. *Restricted geographic range.* Sterile males must disperse to all parts of the range. Small, localized populations are easier to control than large dispersed populations.

Within the last few years, geneticists have begun to devise new genetic control strategies that work by altering DNA or by adding new genes or alleles into the genetic makeup of a pest population. Most of these new tactics are still untested in field applications, but they do offer intriguing possibilities for the future. One such approach involves breeding members of a pest species that have been genetically altered to make them more susceptible to cold winter (or hot summer) temperatures. This type of genetic trait is known as a conditional lethal mutation; it causes mortality only when triggered by extreme environmental conditions. Genetically altered individuals are mass-reared, released into the environment, and allowed to breed with other members of their species. Hopefully, the mutation will spread throughout much of the pest population before the "carriers" are killed by cold (or hot) temperatures.

Another promising approach involves cytoplasmic incompatibility. Although insects lack an immune system, some species do have different "strains" that are not reproductively compatible: the egg cytoplasm of one strain may contain substances that block or inhibit sperm from another strain. By mass-rearing and releasing large numbers of these "incompatible" insects, it may someday be possible to eradicate a pest population and replace it with a more "benign" strain.

Chemical Control

Most people understand that chemical pest control involves the use of chemical substances to kill or disrupt the life cycle of an insect pest. But few people outside the circle of entomology realize just how diverse these compounds are, and in how many different ways they can be used. Although conventional insecticides, the poisons, are still a mainstay of chemical control, they are gradually being superseded by less toxic compounds that disrupt insect development or modify behaviour. Some of these new chemical weapons are much safer for the environment and more species specific than most conventional insecticides.

Semiochemicals — Chemical Control of Behaviour

Much of an insect's behaviour is mediated by chemicals in its environment. By turning these chemicals to our own advantage, it is often possible to attract pests to traps or baits, or repel them from our homes, our crops, or our domestic animals. Behavioral messages are delivered by a wide array of chemical compounds. As a group, these compounds are known as semiochemicals. In some cases, they may facilitate communication between the members of a single species (e.g., pheromones) or between members of different species (e.g., allelochemicals). Functionally, semiochemicals may have a wide range of activity. They may serve as attractants or repellents, they may stimulate or inhibit feeding, they may provoke flight or inhibit it, or they may simply elicit behaviour patterns at inappropriate times.

Attractant pheromones and allelochemicals can be used as lures or baits in a wide variety of insect traps, or they can be mixed together with toxicants to produce an "elixir of death". Protein hydrolysates, for example, serve as feeding attractants for fruit flies (*Rhagoletis* spp.). These chemicals can be applied to sticky traps to improve catch, or combined with an insecticide and sprayed on fruit crops to suppress active infestations. Phenethyl propanoate, eugenol, and geraniol can be mixed in a 3.5:3.5:3 ratio and used as an attractant for Japanese beetles (*Popillia japonica*). These are the active ingredients in the "floral attractant" found in popular bag traps for Japanese beetles. In some cases, chemists have produced synthetic compounds that are even more attractive than naturally occurring chemicals.

Trimedlure, a synthetic substitute for alpha-copaene, is produced commercially as an attractant for the Mediterranean fruit fly (*Ceratitis capitata*). Improved food lures and baits are among the most promising new developments for controlling cockroaches (Blattoidea) in homes and businesses. These are the active ingredients in a new generation of "roach motels" where the insects "check in but don't check out."

Sex pheromones are among the most powerful of chemical attractants. Ever since they were first discovered by A. A. Budenandt in 1959 (from silkworm moths, *Bombyx mori*), these chemicals have aroused great interest because of their potential as pest control agents. During the past 30 years, chemists have identified the sex pheromones for over 300 insect species. Many of these compounds are now sold commercially. In some cases, pheromones are packaged (or encapsulated) in slow-release dispensers (rubber septa, hollow fibres, or rope wicks) that are used as lures in traps of various designs. At low densities, these pheromone traps are a valuable monitoring tool, providing information on the density and distribution of pest populations. At high densities, they can be used for mass trapping sexually active adults (usually males) in efforts to reduce population density and lower a pest's reproductive potential.

Slow-release formulations of sex pheromones can also be used for mating disruption. By increasing the concentration of pheromone in an insect's environment, it may be possible to make everything smell like a prospective mate. Males wear themselves out courting inanimate objects or become habituated to the odor and stop responding to it. This approach, variously known as air permeation or the innundation technique, has shown promise for controlling a number of fruit and vegetable pests, including the codling moth (*Cydia pomonella*), the cabbage looper (*Trichoplusia ni*), the oriental fruit moth (*Grapholita molesta*), and the peachtree borer (*Synanthedon exitiosa*).

Chemical repellents and feeding deterrents are also useful tools for managing insect behaviour. As their name suggests, these compounds cause insects (or other arthropods) to disperse or to discontinue normal feeding behaviour. Repellents such as dimethyl phthalate, benzyl benzoate, and N,N-diethyl-m-toluamide (DEET) have been developed to protect humans from biting flies, ticks, and chiggers (immature trombiculid mites). Other compounds, like di-n-butyl succinate or butoxypolypropylene glycol, are used as fly repellents for cattle.

Moth balls and flakes (paradicholorobenzene or alpha-naphthalene) are placed in drawers and closets to prevent infestation by a variety of insects that feed on stored products and natural fibres. Woolen cloth can be manufactured with colourless dyes (e.g., Eulans and Mitin FF) that bond permanently to the fabric and make it unpalatable to clothes moths and carpet beetles. Wood preservatives, such as pentachlorophenol, act as feeding deterrents to termites and other wood-dwelling insects.

The neem tree, *Azadirachta indica* (Meliaceae) is a promising new source of feeding repellents that may be developed for use on selected non-crop plants. The leaves, twigs, and seeds of this tree, which is grown commercially in India, contain at least 25 biologically active compounds that act as insect repellents, feeding deterrents, or growth regulators. Azadirachtin, the most abundant of these active ingredients, is now commercially available in the United States. Sold under the trade name Margosan-O, this new feeding deterrent has been approved for use on non-food greenhouse crops, ornamentals, and turf.

Chemosterilants — Chemical Control of Reproduction

There are over four hundred chemical substances that are known to cause reproductive sterility in insects. Some of these compounds inhibit ovarian growth and development, while others appear to induce fundamental changes in the chemical structure of nucleic acids (DNA and RNA). These changes (mutations) prevent cell division or obstruct normal embryonic development. Chemosterilants belong to several major chemical groups. These compounds are applied directly to the insect or incorporated into food that serves as a bait. All chemosterilants are extremely hazardous compounds. Their effects are not restricted to insects; they also cause cancer, birth defects, and other mutations in humans and domestic animals. Clearly, these chemicals cannot be dispersed in the environment like other pesticides. Instead, they must be applied under controlled laboratory conditions, usually to insects that are mass reared and released as part of a sterile release program. Although there is much interest in finding a chemosterilant whose effects are limited to insects, no such compound has yet been found.

Insect Growth Regulators — Chemical Control of Development

The enzymes and hormones that regulate developmental processes within an insect's body can sometimes be exploited as chemical control weapons. These compounds, often known as insect growth regulators (IGRs), can be used to stimulate development at inappropriate times or inhibit it at other times. The major groups of IGR compounds include:

1. *Chitin inhibitors.* These chemicals (e.g., diflubenzuron and teflubenzuron) inhibit the molting process (apolysis) by blocking the activity of chitin synthetase, an enzyme needed by epidermal cells when constructing a new exoskeleton. Because of this mode of action, chitin inhibitors are highly specific to arthropods. They act rather slowly (2-5 days), but eventually disrupt any process that involves construction of new cuticle (e.g., molting, hatching, pupation). They are most effective when used against the immature stages of a pest. Diflubenzuron, currently registered under the trade name Dimilin, is used for controlling gypsy moths, boll weevils, and various other pests.
2. *Molting Hormone Analogues.* Ecdysteroids stimulate the molting process by mimicking the action of molting hormone. Applied to the surface of an insect's body or incorporated into its food, these compounds work by initiating premature ecdysis during the immature stages of development. Ecdysteroid-like compounds have been found in some plants where they evidently serve as a defense against insect herbivores. But despite their potential as insect growth regulators, the ecdysteroids have never been developed into commercial products. Their chemical structural is similar to that of human reproductive hormones (estrogen, progesterone, and testosterone), and like many other steroid compounds, they have the potential to cause cancer and birth defects.

3. *Juvenile Hormone Analogues.* Juvenile hormone (JH) and related compounds act as insect growth regulators by inhibiting the developmental changes associated with embryogenesis, morphogenesis, and reproduction. During normal development, JH levels are elevated in larvae (or nymphs) and decrease prior to pupation (or adult eclosion). Contact exposure to JH analogues during the egg stage or after the last larval (or nymphal) molt can inhibit development, delay maturation, and eventually result in death. Since the onset of mortality is usually quite slow (days to weeks), JH analogues have limited utility in agriculture. But several compounds (e.g., hydroprene, kinoprene, and methoprene) have been successfully incorporated into household products for controlling ants, fleas, and other household pests.
4. *Anti-juvenile Hormones.* These unique compounds (the precocenes) were first isolated in 1976 from a common houseplant (*Aegeratum houstonianum*). Precocenes are cytotoxic agents. They become activated by enzymes in the insect's corpora allata, selectively destroying these glands, and preventing all subsequent production of juvenile hormone. In immature insects, exposure to anti-JH compounds may result in premature (precocious) development of adult structures or behaviors. In adults, precocenes can cause sterility because the presence of juvenile hormone is necessary for normal production of eggs and sperm. Anti-JH compounds seem to be most effective against Hemipterans. Despite their unique mode of action, these IGRs have never been developed into commercial products because they break down too rapidly in the presence of oxygen.

Conventional Insecticides — The Killer Chemicals

Conventional insecticides are among the most popular chemical control agents because they are readily available, rapid acting, and highly reliable. A single application may control several different pest species and usually forms a persistent residue that continues to kill insects for hours or even days after application. Because of their convenience and effectiveness, insecticides quickly became standard practice for pest control during the 1960's and 1970's. Overuse, misuse, and abuse of these chemicals have led to widespread criticism of chemical control and, in a few cases, resulted in long-term environmental consequences.

The effectiveness of an insecticide usually depends on when and where the pest encounters it. Most insecticides are absorbed directly through an insect's exoskeleton. These compounds are known as contact poisons because they are effective "on contact". Other insecticides act as fumigants. They are released in the vapour state (as gases) and enter the insect's body through its tracheal system. Fumigants are most effective when they are used in an enclosed area such as a greenhouse, a warehouse, or a grain bin. Still other compounds must be ingested before they have an effect. These are known as stomach poisons. They often work more slowly than fumigants or contact poisons, but they are still useful for certain types of pest control in homes and businesses.

Systemic insecticides are a special type of stomach poison. These compounds are absorbed by the tissues of a plant (or animal) without ill effects. Insect pests ingest the insecticide when they feed on the treated organism. Systemic insecticides are sometimes included in the diets of domestic animals to protect them from internal parasites (e.g., cattle grubs and other bot flies). Plant systemics can be incorporated into the soil around ornamentals or bedding plants. The insecticides are absorbed by the roots and translocated to leaves, stems, and flowers. If the insect that feeds on a treated plant doesn't acquire a lethal dose of insecticide, it may at least be deterred from further feeding. Although systemic insecticides are commonly applied to horticultural plantings, they are not as useful for many food crops because the insecticide remains in the food after harvest.

Insecticides contain one or more active ingredients that serve as toxicants (poisons). In their purest form (technical grade), these chemicals may be too toxic, too unstable, or too volatile to be handled or applied safely. Therefore, technical grade insecticide is always mixed with other compounds, known as adjuvants, in order to improve the performance, safety, or handling characteristics of a commercial product. These mixtures (technical grade insecticide plus adjuvants) are known as formulations. Almost anything could be an adjuvant: pumice, ground walnut shells, buffalo gourd root powder, vegetable oil, etc. These compounds are usually listed on the label as "inert ingredients", but they are certainly not inactive. Many adjuvants are proprietary products, protected by patents and closely guarded as industrial secrets. They may represent 90-95% of the total volume of a commercial formulation. An insecticide can be manufactured in a variety of formulations, each tailored for specific applications. Some of the more common formulations include:

Granules (G) or pellets (P) — Coarse particles (e.g., clay, ground corn cobs, or walnut shells) can serve as a carrier for certain types of insecticides. The toxicant slowly leaches out of the carrier, minimizing its movement within the ecosystem and maximizing its active life (persistence). Granular formulations are commonly used for controlling soil-dwelling insects and in systemic insecticides that are applied around the base of plants. Since granular formulations do not blow or drift, they are considered relatively safe from the standpoint of accidental human exposure.

Dusts (D) — These dry powders are usually formulated with inert particles of ash, chalk, talc, or clay. They are designed to be sprinkled or blown onto target surfaces. The powder sticks to feet, legs, and other body parts of passing insects, and the toxicant is eventually ingested when the insect cleans itself. Insecticides that act as stomach poisons are often formulated as dusts. These formulations are not always suitable for outdoor applications because they have a tendency to drift (blow away with the wind).

Soluble powders (SP) or wettable powders (WP) — Dry formulations that are mixed in water to form homogeneous spray solutions. A soluble powder dissolves completely in water, whereas a wettable powder contains an emulsifier that produces a uniform suspension (colloid). Many foliar insecticides are formulated as wettable (or soluble) powders because they are easy to transport and generally have a long shelf life.

Emulsifiable concentrates (EC) — These liquid formulations contain the toxicant(s) and an emulsifier dissolved in organic solvent. The concentrate is diluted with a large volume of water to produce the final spray mixture. Some foliar insecticides are formulated as emulsifiable concentrates. Unlike most wettable powders, they do not leave a visible residue on fruits and vegetables. Sensitive plants, however, may be injured by organic solvents in the mixture. Emulsifiable concentrates are also commonly used in sprays for urban and industrial pests.

Aerosols (A) — Insecticides that are formulated together with a solvent may be pre-packaged in pressurized spray cans or sold unpressurized for use in special fogging machines. Spray cans are relatively expensive (per pound of active ingredient) but they are convenient, easy to store, and have a long shelf life. Commercial foggers are typically used indoors (e.g., greenhouses and warehouses) or for control of biting flies in community-wide pest control operations.

Ultralow-Volume Concentrates (ULV) — These highly concentrated formulations (more than 8 pounds of active ingredient per gallon) are designed to be used in specialized spray equipment that atomizes the concentrate droplets. ULV spray equipment is used by most aerial applicators (airplane sprayers) who treat forested lands or large agricultural acreages. Despite their many advantages, conventional insecticides are not ideal pest control agents. Indeed, one of their greatest strengths, broad-spectrum activity, is also one of their greatest weaknesses. While it is certainly an advantage to control multiple pest species with a single chemical treatment, the non-specificity of most conventional insecticides poses a serious threat to non-target organisms in the environment. High mortality among natural enemies can have an enduring impact on the ecological balance of any community. In the absence of biocontrol agents, more insecticide applications may be the only recourse available to stop pest resurgence. Once we step onto this "insecticide treadmill", it can be very difficult to get off.

The non-target effects of insecticides are not limited to natural enemies. Most of the organochlorine compounds (e.g., DDT) have been banned from use in the United States because of their indirect effects on reproduction in birds of prey (e.g., eagles, ospreys, condors, etc.). These environmentally stable compounds are highly soluble in lipids, making it possible for them to accumulate in the body fat of non-target organisms. Predators, particularly those at the top of a food chain, amass pesticide concentrations many times greater than anywhere else in the environment. This process, known as bioaccumulation (or biomagnification), was responsible for high levels of DDT and related compounds in birds of prey during the 1960's and 1970's. These pesticides did not injure the birds themselves, but caused thinning of their egg shells and high rates of breakage during incubation. Most organochlorine insecticides were banned during the 1970's and 1980's and many of the threatened bird species are now recovering from the brink of extinction.

In some cases, sub-lethal concentrations of an insecticide can stimulate rather than suppress the growth of a pest population. This phenomenon, known as hormoligosis, has been observed in a number of pest species, including twospotted spider mites (*Tetranychus urticae*), western corn rootworms (*Diabrotica virgifera*), and brown planthoppers (*Nilaparvata lugens*). Low doses of pesticide seem to improve the nutritional quality of host plants, thereby increasing a pest's reproductive potential or decreasing its time of development.

And finally, there is the problem of resistance to insecticides. Insects are among the most adaptable organisms on the face of the earth. For the past 400 million years they have managed to survive by adjusting to changes in their environment, so it should come as no surprise that they can also adapt to chemical pesticides. Resistance has increased exponentially since the the late 1940's, and today, there are over 500 pest species that exhibit some level of resistance to at least one type of insecticide.

Insects may become resistant to insecticides in several ways. Biochemical resistance usually involves changes in the metabolic pathways that insects normally use to break down plant defenses and other environmental toxins. This detoxication is facilitated by enzymes (esterases, hydrolases, transferases, and oxidases) that change the chemical structure of toxicants before they cause physical harm. Physiological resistance involves functional changes in basic life processes that alter the way toxicants interact with the body. Some German cockroaches (*Blatella germanica*), for example, have become resistant to carbaryl (a carbamate insecticide) as a result of genetic changes in the permeability of their cuticle. Behavioral resistance may occur as a result of any innate change in behaviour that reduces an insect's probability of encountering a toxicant. In some parts of Panama, the mosquitoes that vector malaria (*Anopheles albimanus*) have become hypersensitive to certain insecticides. They manage to escape lethal doses of insecticide by avoiding enclosed areas and refusing to land on treated surfaces.

When an insect population becomes resistant to one insecticide, it may also become less susceptible to other toxicants in the same chemical family. This phenomenon, known as class resistance, is a common problem in all major groups of insecticides (organochlorines, organophosphates, carbamates, and synthetic pyrethroids). When an insecticide loses its effectiveness because of pest resistance, users typically replace it with another compound from a different chemical group. But resistance to one group of compounds does not prevent subsequent development of resistance to compounds in other chemical groups. Over the years, some pests have developed multiple resistance to nearly every insecticide that has ever been used to fight them. Colorado potato beetles, *Leptinotarsa decemlineata*, on Long Island, N.Y. are probably the best example of a population that has developed multiple resistance. It may never be possible to prevent insects from developing resistance, but there are ways to manage these chemicals that will prolong their useful life.

Integrated Control

Just as ground, air, and naval forces are integrated to achieve military objectives, the tactical weapons of pest control can also be integrated to achieve more effective management of pest populations. Development of resistance, effects on non-target organisms, and damage to the environment can all be minimized with selective and judicious use of multi-faceted control tactics. This approach, commonly known as integrated control, requires an understanding of ecological principles as well as a thorough knowledge of the pest's life history and population dynamics. Integrated pest control is not a new concept. It was commonly practiced in the years before synthetic organic insecticides became widely available. But the old ways were largely abandoned after World War II because chemical weapons were so effective, convenient, and inexpensive. Once we recognized the dangers of over-dependence on a single control strategy, the principles of integrated pest control gained renewed acceptance. Today, integrated pest control forms the foundation of Integrated Pest Management programs (IPM) that take a comprehensive and multi-disciplinary approach to solving pest problems. Insects, weeds, plant diseases, and even some vertebrate pests (e.g., birds and rodents) are included under the IPM umbrella. These programs emphasize management rather than eradication. They take a broad ecological approach to pest problems, focusing on all members of a pest complex in an effort to identify the optimum combination of control tactics that will reduce pest populations below economic thresholds and maintain these levels with the least possible impact on the rest of the environment. This approach, often called biorational pest control, relies heavily on cultural and biological tactics that are supplemented with carefully timed applications of highly selective chemical weapons.

The complexity of modern IPM programs will continue to increase as we add more knowledge about pests, develop new management tactics, and learn how to optimize existing control strategies. Mathematical models of population growth rates and plant-pest interactions are being developed and incorporated into computer programs that help us assimilate and interpret data from the many different variables that affect pest population dynamics. These computer programs, usually known as expert systems, are yet another tool in our expanding stockpile of pest control technology.

Integrated pest management is also a major component of Sustainable Agriculture, a holistic approach to modern farming that encompasses not only pest control, but also conservation of soil and water resources, and a wide range of other management practices (e.g., composting, mulching, green manure, waste disposal, etc.). The goal of sustainable agriculture is to satisfy human needs for food and fiber while preserving nonrenewable resources, protecting environmental quality for future generations, and safeguarding the profitability and long-term viability of commercial agriculture.

11

Integrated Pest Management

In agriculture, integrated pest management (IPM) is a pest control strategy that uses a variety of complementary strategies including: mechanical devices, physical devices, genetic, biological, cultural management, and chemical management.

These methods are done in three stages: prevention, observation, and intervention. It is an ecological approach with a main goal of significantly reducing or eliminating the use of pesticides while at the same time managing pest populations at an acceptable level.

History of IPM

Shortly after World War II, when synthetic insecticides became widely available, entomologists in California developed the concept of "supervised insect control."

Around the same time, entomologists in cotton-belt states such as Arkansas were advocating a similar approach. Under this scheme, insect control was "supervised" by qualified entomologists, and insecticide applications were based on conclusions reached from periodic monitoring of pest and natural-enemy populations.

This was viewed as an alternative to calendar-based insecticide programs. Supervised control was based on a sound knowledge of the ecology and analysis of projected trends in pest and natural-enemy populations.

Supervised control formed much of the conceptual basis for the "integrated control" that University of California entomologists articulated in the 1950s. Integrated control sought to identify the best mix of chemical and biological controls for a given insect pest. Chemical insecticides were to be used in manner least disruptive to biological control. The term "integrated" was thus synonymous with "compatible." Chemical controls were to be applied only after regular monitoring indicated that a pest population had reached a level (the economic threshold) that required treatment to prevent the population from reaching a level (the economic injury level) at which economic losses would exceed the cost of the artificial control measures. IPM extended the concept of integrated control to all classes of pests and was expanded to include tactics other than just chemical and biological controls. Artificial controls such as pesticides were to be applied as in integrated control, but these now had to be compatible with control tactics for all classes of pests. Other tactics, such as host-plant resistance and cultural manipulations, became part of the IPM arsenal. IPM added the multidisciplinary element, involving entomologists, plant pathologists, nematologists, and weed scientists.

In the United States, IPM was formulated into national policy in February 1972 when President Nixon directed federal agencies to take steps to advance the concept and application of IPM in all relevant sectors. In 1979, President Carter established an interagency IPM Coordinating Committee to ensure development and implementation of IPM practices.

How IPM Works

An IPM regime can be quite simple or sophisticated. Historically, the main focus of IPM programs was on agricultural insect pests. Although originally developed for agricultural pest management, IPM programs are now developed to encompass diseases, weeds, and other pests that interfere with the management objectives of sites such as residential and commercial structures, lawn and turf areas, and home and community gardens.

An IPM system is designed around six basic components: The US Environmental Protection Agency has a useful set of IPM principles.

1. *Acceptable Pest Levels:* The emphasis is on *control*, not *eradication*. IPM holds that wiping out an entire pest population is often impossible, and the attempt can be more costly, environmentally unsafe, and frequently unachievable. IPM programs first work to establish acceptable pest levels, called action thresholds, and apply controls if those thresholds are crossed. These thresholds are pest and site specific, meaning

that it may be acceptable at one site to have a weed such as white clover, but at another site it may not be acceptable. This stops the pest gaining resistance to chemicals produced by the plant or applied to the crops. If many of the pests are killed then any that have resistance to the chemical will rapidly reproduce forming a resistant population. By not killing all the pests there are some un-resistant pests left that will dilute any resistant genes that appear.

2. *Preventive Cultural Practices:* Selecting varieties best for local growing conditions, and maintaining healthy crops, is the first line of defense, together with plant quarantine and 'cultural techniques' such as crop sanitation (*e.g.* removal of diseased plants to prevent spread of infection).
3. *Monitoring:* Regular observation is the cornerstone of IPM. Observation is broken into two steps, first; inspection and second; identification. Visual inspection, insect and spore traps, and other measurement methods and monitoring tools are used to monitor pest levels. Accurate pest identification is critical to a successful IPM program. Record-keeping is essential, as is a thorough knowledge of the behaviour and reproductive cycles of target pests. Since insects are cold-blooded, their physical development is dependent on the temperature of their environment. Many insects have had their development cycles modeled in terms of degree days. Monitor the degree days of an environment to determine when is the optimal time for a specific insect's outbreak.
4. *Mechanical Controls:* Should a pest reach an unacceptable level, mechanical methods are the first options to consider. They include simple hand-picking, erecting insect barriers, using traps, vacuuming, and tillage to disrupt breeding.
5. *Biological Controls:* Natural biological processes and materials can provide control, with minimal environmental impact, and often at low cost. The main focus here is on promoting beneficial insects that eat target pests. Biological insecticides, derived from naturally occurring microorganisms (*e.g.*: *Bt*, entomo-pathogenic fungi and entomopathogenic nematodes), also fit in this category.
6. *Chemical Controls:* Synthetic pesticides are generally only used as required and often only at specific times in a pests life cycle. Many of the newer pesticide groups are derived from plants or naturally occurring substances (*e.g.*: nicotine, pyrethrum and insect juvenile hormone analogues), and further 'biology-based' or 'ecological' techniques are under evaluation.

IPM is applicable to all types of agriculture and sites such as residential and commercial structures, lawn and turf areas, and home and community gardens. Reliance on knowledge, experience, observation, and integration of multiple techniques makes IPM a perfect fit for organic farming (the synthetic chemical option is simply not considered). For large-scale, chemical-based farms, IPM can reduce human and environmental exposure to hazardous chemicals, and potentially lower overall costs of pesticide application material and labour.

1. *Proper Identification of Pest:* What is it? Cases of mistaken identity may result in ineffective actions. If plant damage due to over-watering are mistaken for a fungal infection, a spray may be used needlessly and the plant still dies.
2. *Learn Pest and Host Life Cycle and Biology:* At the time you see a pest, it may be too late to do much about it except maybe spray with a pesticide. Often, there is another stage of the life cycle that is susceptible to preventative actions. For example, weeds reproducing from last year's seed can be prevented with mulches. Also, learning what a pest needs to survive allows you to remove these.
3. *Monitor or Sample Environment for Pest Population:*How many are here? Preventative actions must be taken at the correct time if they are to be effective. For this reason, once you have correctly identified the pest, you begin monitoring BEFORE it becomes a problem. For example, in school cafeterias where roaches may be expected to appear, sticky traps are set out before school starts. Traps are checked at regular intervals so you can see them right away and do something before they get out of hand. Some of the things you might want to monitor about pest populations include: Is the pest present/absent? What is the distribution-all over or only in certain spots? Is the pest population increasing or decreasing?
4. *Establish Action Threshold (economic, health or aesthetic):* How many are too many? In some cases, a certain number of pests can be tolerated. Soybeans are quite tolerant of defoliation, so if you have only a few caterpillars in the field and their population is not increasing dramatically, there is no need to do anything. Conversely, there is a point at which you MUST do something. For the farmer, that point is the one at which the cost of damage by the pest is MORE than the cost of control. This is an economic threshold. Tolerance of pests varies also by whether or not they are a health hazard (low

tolerance) or merely a cosmetic damage (high tolerance in a non-commercial situation). Personal tolerances also vary-many people dislike any insect; some people cannot tolerate dandelions in their yards. Different sites may also have varying requirements based on specific areas. White clover may be perfectly acceptable on the sides of a tee box on a golf course, but unacceptable in the fairway where it could cause confusion in the field of play.

5. Choose an appropriate combination of management tactics For any pest situation, there will be several options to consider. Options include, mechanical or physical control, cultural controls, biological controls and chemical controls. Mechanical or physical controls include picking pests off plants, or using netting or other material to exclude pests such as birds from grapes or rodents from structures. Cultural controls include keeping an area free of conducive conditions by removing or storing waste properly, removing diseased areas of plants properly. Biological controls can be support either through conservation of natural predators or augmentation of natural predators. Augmentative control includes the introduction of naturally occurring predators at either an inundative or inoculative level. An inundative release would be one that seeks to inundate a site with a pest's predator to impact the pest population. An inoculative release would be a smaller number of pest predators to supplement the natural population and provide ongoing control. Chemical controls would include horticultural oils or the application of pesticides such as insecticides and herbicides. A Green Pest Management IPM program would use pesticides derived from plants, such as botanicals, or other naturally occurring materials.
6. *Evaluate Results:* How did it work? Evaluation is often one of the most important steps. This is the process to review an IPM program and the results it generated. Asking the following questions is useful: Did your actions have the desired effect? Was the pest prevented or managed to your satisfaction? Was the method itself satisfactory? Were there any unintended side effects? What will you do in the future for this pest situation? Understanding the effectiveness of the IPM program allows the site manager to make modifications to the IPM plan prior to pests reaching the action threshold and requiring action again.

Integrated Pest Management (IPM) Principles

What is IPM?

Integrated Pest Management (IPM) is an effective and environmentally sensitive approach to pest management that relies on a combination of common-sense practices. IPM programs use current, comprehensive information on the life cycles of pests and their interaction with the environment. This information, in combination with available pest control methods, is used to manage pest damage by the most economical means, and with the least possible hazard to people, property, and the environment. The IPM approach can be applied to both agricultural and non-agricultural settings, such as the home, garden, and workplace. IPM takes advantage of all appropriate pest management options including, but not limited to, the judicious use of pesticides. In contrast, *organic* food production applies many of the same concepts as IPM but limits the use of pesticides to those that are produced from natural sources, as opposed to synthetic chemicals.

How do IPM Programs Work?

IPM is not a single pest control method but, rather, a series of pest management evaluations, decisions and controls. In practicing IPM, growers who are aware of the potential for pest infestation follow a four-tiered approach. The four steps include:

Set Action Thresholds: Before taking any pest control action, IPM first sets an action threshold, a point at which pest populations or environmental conditions indicate that pest control action must be taken. Sighting a single pest does not always mean control is needed. The level at which pests will either become an economic threat is critical to guide future pest control decisions.

Monitor and Identify Pests: Not all insects, weeds, and other living organisms require control. Many organisms are innocuous, and some are even beneficial. IPM programs work to monitor for pests and identify them accurately, so that appropriate control decisions can be made in conjunction with action thresholds. This monitoring and identification removes the possibility that pesticides will be used when they are not really needed or that the wrong kind of pesticide will be used.

Prevention: As a first line of pest control, IPM programs work to manage the crop, lawn, or indoor space to prevent pests from becoming a threat. In an agricultural crop, this may mean using cultural methods, such as rotating between different crops, selecting pest-resistant varieties, and planting pest-free rootstock. These control methods can be very effective and cost-efficient and present little to no risk to people or the environment.

Control: Once monitoring, identification, and action thresholds indicate that pest control is required, and preventive methods are no longer effective or available, IPM programs then evaluate the proper control method both for effectiveness and risk. Effective, less *risky* pest controls are chosen first, including highly targeted chemicals, such as pheromones to disrupt pest mating, or mechanical control, such as trapping or weeding. If further monitoring, identifications and action thresholds indicate that less risky controls are not working, then additional pest control methods would be employed, such as targeted spraying of pesticides. Broadcast spraying of non-specific pesticides is a last resort.

Do Most Growers use IPM?

With these steps, IPM is best described as a continuum. Many, if not most, agricultural growers identify their pests before spraying. A smaller subset of growers use less risky pesticides such as pheromones. All of these growers are on the IPM continuum. The goal is to move growers further along the continuum to using all appropriate IPM techniques.

How do you know if the Food you Buy is Grown Using IPM

In most cases, food grown using IPM practices is not identified in the marketplace like *organic* food. There is no national certification for growers using IPM, as the United States Department of Agriculture has developed for organic foods. Since IPM is a complex pest control process, not merely a series of practices, it is impossible to use one IPM definition for all foods and all areas of the country. Many individual commodity growers, for such crop as potatoes and strawberries, are working to define what IPM means for their crop and region, and IPM-labelled foods are available in limited areas. With definitions, growers could begin to market more of their products as *IPM-Grown*, giving consumers another choice in their food purchases.

If I grow my own fruits and vegetables, can I practice IPM in my garden?

Yes, the same principles used by large farms can be applied to your own garden by following the four-tiered approach outlined above. For more specific information on practicing IPM in your garden, you can contact your state Extension Services for the services of a Master Gardener.

Major Insect Pest Management of Sugarcane Crop

Sugarcane is a long duration crop of 10-18 months and therefore is liable to be attacked by a number of insect pests and diseases. According to an estimate, sugarcane production declines by 20.0 and 19.0 % by insect pests and diseases respectively. To increase the crop productivity, management of insect-pest and diseases is of great significance. Due to diversity in agro-ecological conditions the importance of insect pests and disease varies and therefore, management strategy should be adopted accordingly. Sugarcane is infested by about 288 insects of which nearly two dozen causes heavy losses to the quality as well as quantity of the crop. The scenario of insect pests and diseases varies in sub-tropical and tropical belt of sugarcane. Top borer and stalk borer are found pre-dominantly in sub-tropical areas whereas internodes borer and early shoot borer and among disease rust & eye spot are prevalent in tropical region

Several management strategies have been developed as a result of research and development work. In order to save environment from chemical pollution, use of bio-control has been given utmost attention. The management technologies have been integrated as per need for increasing the efficiency.

Economic Threshold Level (ETL) of Sugarcane Pests

ETL is the density at which control measures should be determined to prevent an increase pest population from reaching economic injury level. This helps to check the pests before they can cause notable damage to sugarcane crop.

Pests of Plants: Sugarcane Stem-borer

Chilo Terrenellus

Profile New Guinea is the 'home' of sugarcane. Commercial sugarcane grown in Australia (and other parts of the world) has been developed from hybrids. These canes are very susceptible to damage by pests occurring in New Guinea and other countries. There are several species of sugarcane stem-borers in New Guinea. The common species in sugarcanes in the coastal villages of the Western Province is *Chilo terrenellus*. Stem-borers cause extensive damage in commercial plantings of sugarcane in New Guinea. The caterpillar feeds within the stem of growing canes and may cause enough damage to kill the growing point, resulting in browning of the upper leaves. Other signs are holes in the stem and frass caught in the axils of the lower leaves.

Some species also feed in the stems of rice, sorghum and corn. Damage to the upper stems of corn plants can result in loss of the male flower followed by poor pollination of the cobs.

Distribution: These pests are widespread in New Guinea and are found intermittently on the nearby islands of Saibai and Dauan in Torres Strait. Damage levels as high as 70-80% have been observed in sugarcane in the Western Province of New Guinea.

Threat: The commercial sugarcane species are softer than the 'wild' canes and have higher sugar content. If these pests were to enter our commercial growing areas they would have a severe impact on production and be very difficult to control.

Keep a Top Watch Don't bring sugarcane from New Guinea into Torres Strait. Don't move sugarcane from the northern islands of Torres Strait. If you see sugarcane with holes tunnelled in the stem report it to your local Quarantine officer. There are also State regulations on the movement of sugarcane. Please observe these regulations to help stop the movement of pests and disease of sugarcane.

Disease Management

- Currently there are about 50 varieties of sugarcane under intensive cultivation in India.
- Over 100 fungi, 10 bacteria, 10 viruses and about 50 species of nematodes are pests of sugarcane in different parts of the world.
- Wherever the crop is grown intensively virulent forms of certain pathogens are chronic.
- When a given pathogen becomes severe enough to get the upper hand, the farmers replace the susceptible variety with resistant or tolerant varieties.
- Thus, in many parts of the world, including India, there are continuous, forced replacements of sugarcane varieties.
- There are over 30 diseases of sugarcane in India, some are of major importance because they are widespread and cause severe losses in yield.
- Red rot is one of the major diseases of sugarcane found in many areas of the world. In India it has caused extensive damage in recent past and remains endemic in severe form in some parts.
- During 1939-1942 it caused heavy losses in Bihar and Uttar Pradesh.

- The disease has been studied in detail during the past half century, as evidenced by the voluminous literature on it.
- Butler and his associates in India investigated the disease in great detail during 1914-1918. This led to the work on breeding for disease resistant varieties to control the disease.
- A lead in this activity was by The Sugarcane Breeding Institute, Coimbatore, South India.

Symptoms

- The first symptom of red rot in the field is discolouration of the young leaves. The margins and tips of the leaves wither and the leaves droop.
- The discolouration and withering continues from the tip to the leaf base until the whole crown withers and the plant dies, within 4 to 8 days.
- In a single stool, most of the stalks may wither almost simultaneously.
- As the disease advances the entire stem rots and the central tissues become pithy.
- The tissues are reddened throughout the basal portion, especially the vascular bundles, which are intensely red; there may be cross-wise white patches interrupting the reddened tissues.
- The internodes may shrink and when such canes are split open large cavities may be found in the centre and the pithy tissues may appear brown.
- Often a profuse whitish growth of the fungal mycelium may be found in the brown background of the host tissue.
- In some cases, black, minute, velvety bodies, representing the acervuli of tire fungus, may also be seen
- Since reddening is a common symptom of other diseases of sugarcane, the white patch symptom is an important diagnostic characteristic of red rot.
- When a diseased plant is open, a characteristic becomes evident.
- In the infected plants the leaves may show symptoms in the form of dark red lesions in the mid-rib, which may elongate, turning blood-red with dark margins and later on with straw-coloured centres.
- In the older lesions minute black dots, representing the acervuli, can be seen. Often the infected leaves may break at the lesions and hand down

- The fungal mycelium is present inter-cellularly, mainly in the parenchymatous cells of the path.
- The hyphae are thin, hyaline, septate, profusely branching and contain oil droplets.
- The acervuli, which are characteristic of the genus, are formed on the surface of the rind, in the leaf mid-rib and sometimes in the pith region. They are minute, black and velvety.
- The compact hymenial layer produces the long, rigid, bristle-like setae, which are septate, dark brown at the base and lighter towards the tip.
- Inter-spaced between the setae are numerous club-shaped to linear, hyaline, single-celled conidiophores, measuring approximately 20 X 8 m.
- The conidia, which are borne singly oil the conidiophores, are single celled, thin walled, hyaline, falcate, granular and guttulate, measuring 16-48 x 4-8m In artificial cultures terminal and intercalary, thick-walled, greenish-black chlamydospores are often produced.
- The conidia germinate readily to produce single germ tubes at the tips of which an elliptical, ovoid or irregular, appresoria are formed.
- The perfect stage of the fungus has been observed both in nature and in culture by several authors.

Disease Cycle

- The planting material, viz., the setts, may harbour the fungus and thus perptuate the disease from season to season.
- The fungus may also persist in the soil oil diseased clumps and dry leaves left in the field after harvest.
- The primary infection, however, appears to be mainly from infected setts.
- Secondary spread in the field may be through irrigation water, cultivation tools anti implements and wind-borne inoculum.
- If the conidia settle on the leaves they may germinate and invade the leaves through various types of wounds including the splitting of the mid-rib so common in many varieties.
- Stem infection may take place through insect bores and root primordia.
- The soil-borne fungus may also enter the healthy setts through cut-ends, and cause early infection of the shoots.

- The prevalence of several pathogenic strains of the fungus has been reported from many countries, including India.
- In general, light-coloured physiologic races sporulate readily and are more pathogenic than the dark-coloured strains that sporulate sparingly.
- Though the perfect stage of the fungus has been observed in nature, the role of ascospores in the disease cycle is not understood.

Control

- Red rot management in sugarcane has become an important issue in all sugarcane areas. While early stage detection may not be quite easy, during later stages, the cane breaks down.
- The canes are to be split open lengthwise to see dull red tissue, interrupted by white patches across the stalk. These patches are characteristic of red rot of sugarcane. Monsoon period enables faster disease spread and drying of the crop.
- The best remedy for avoiding this fungal disease is to cultivate only resistant sugarcane varieties that have been released for cultivation in different sugarcane growing states. Phytosanitation, being the key to manage this disease, stringent domestic quarantine measures to prevent movement of cane setts from endemic areas to new areas has to be enforced by all concerned agencies including sugar mills.
- In recent times, the disease has spread into many states.
- Following good cultural practices such as clearing fields of excessive trash and ensuring efficient drainage.
- Healthy setts only are to be planted to avoid poor plant stand due to rotting.
- Affected fields should be isolated through bunding to prevent movement of water to adjacent fields.
- Rattooning of infected fields should be strictly avoided.
- Crop rotation in the affected fields could reduce disease inoculum.
- Hot water treatment of setts before planting at 52 degree C for 30 minutes is also recommended.
- Planting fungicide treated seed setts is to be followed as a general practice to prevent introduction of the pathogen into sugarcane fields.
- *Chemical Control:* Sett treatment with 0.25% Thiophanate methyl, a systemic fungicide will protect the setts against the primary infection of red rot for 90 days.

- In addition to the various sanitary precautions mentioned above, red rot may also be controlled by growing resistant or tolerant varieties. Inter-generic and inter-specific crosses with Saccharum spp., in the latter case using S.spopntaneum and S. robustum with S. officinarum have yielded many economically important cane varieties, some of which are highly resistant, or at least tolerant, to red rot.

Red-Stripe

Causal Organism: Xanthomonas rubrilineans

Class: Schizomycetes

Order: Pseudomonadales

Family: Pseudomonadaceae

- This is another bacterial disease of sugarcane which has a world-wide distribution.
- It has been known since 1893 in Java and since 1903 in Australia, but in India only since 1933, when it was reported by Desai, and also McRae.
- In 1960 Rangaswami made detailed studies on the disease and its causal agent.

Symptoms

- Red stripe is characterized by the appearance on the leaves of chlorotic lesions carrying dark red stripes 0.5-1.0 mm in breadth and several mm in length, either distributed all over the blade, or concentrated in the middle
- Several of them may coalesce to cover large areas of the leaf blade, and to cause wilting and drying of the leaves.
- Whitish flakes occur on the lower surface of the leaf, corresponding to the red lesions on the upper surface.
- These flakes are the dry bacterial ooze. When young shoots are affected, shoot or top rot may result.
- The growing points of the shoot are yellow and later reddish with dark brown stripes on the shoots.
- The rotting may commence from the tip and spread downwards.
- If the affected plants are cut by splitting the shoot downwards, dark red dis-colouration of the tissues may be seen.
- In the affected canes cavities may form in the pith region, and the vascular bundles are distinct because of the dark red dis-colouration.

- The diseased and rotting shoots can be easily pulled out and separated from the plant.

Disease Cycle

- The disease spreads in the field by wind and rain, and by cutting, as the basal stem from which the setts are taken is mostly free from the bacterial infection.
- The bacterial ooze form the infected leaves and shoots dries up to form a thin crust on the surface.
- When dry, the bacterial cells spread freely by wind.
- The cells falling on the host plants, enter them through natural openings or wounds and establish themselves in the various tissues, including the xylem.
- Infected parenchymatous cells may collapse and normal functioning of the plant parts may fail.
- Several grasses, including ragi and bajra, have been reported to be infected by the bacteria.
- These hosts may also play a role in the perpetuation and spread of the pathogen.

Control

- This is a difficult disease of sugarcane to control.
- When-ever the disease is noticed, the affected plants should be removed and burnt.
- Such systematic destruction of the affected plants reduces the disease incidence.
- Growing resistant varieties is, however,the best method of control.

Sett Rot or Pine Apple Disease

Causal Organism: Ceralocystis paradoxa, Ceratostomella paradoxa

Class: Ascomycetes

Order: Sphaeriales

Family: Ceratostomataceae.

- This disease of sugarcane is common in India and in most of the sugarcane areas of the world including Hawaii, mainland United States, Mexico, West Indies, parts of Africa, and South America.
- In India it is found in all the sugarcane tracts.

Symptoms

- The disease primarily affects the sugarcane setts.
- When diseased setts are planted they may rot before germination, or the shoots may die after reaching a height of about 6-12 inches.
- If infected shoots survive, they are very much stunted and chlorotic.
- Eventually the leaves may wither and the shoots wilt.
- If the affected shoots and setts are examined the central portion of the shoots will be seen discoloured red and the contents of the sett rotting.
- A characteristic pineapple smell is associated with the rotting and hence the name.
- The fungus produces both microconidia and macroconida.
- The microconidia are hyaline when young but become almost black at maturity.
- They are thin walled aμm the chains of microconidia are produced endogenousnd cylindrilorm, measuring 10-15 μm x 3.5-5.0μm in the conidiophores and are pushed out in succession.
- The conidiophores are linear, thin-walled, about 100μm in length and are formed terminally on short basal cells.
- The macroconida are produced in chains of up to 20μm short, lateral conidiophores, which measure 20-80 x 4 μm. They are spherical, elliptical, truncale or sometimes pyriform and are hyaline to olive-green or black, measuring 16-19 x 10-12 μm.
- The fungus produces the ascigerous of perfect stage in culture media and at times on the host tissues.
- The perithecia are gregarious and flask shaped, with a long narrow beak.
- The basal portion measures 200-350 μm in diameter and the beak up to about 1200μm in length and 30-40μm in diameter.
- The asci are clavate, and the ascospores are convex to elliptical and measure 7-10 x 2.5-4 μm.

Disease Cycle

- The fungus is soil-borne, entering the setts through the cut ends.
- Inside the sett it spreads rapidly through the parenchymatoous tissues and causes sett rot.

- The buds on the setts may also be affected and secondary organisms may hasten the decay.
- Occasionally the infected setts may also be responsible for the spread of the pathogen.
- The micro and macroconidia may live saprophytically in soil for several years and infect the setts when planted.
- Insects also play a part in the dissemination of the pathogen.
- The specific role of the ascigerous stage in perpetuating the fungus is not under stood.
- The same fungus also infects banana, papaya, cocoa, coffee, mango, pineapple, arecanut, coconut, palmhyra, date palm and several other hosts.

Control

- Healthy setts should be obtained from disease-free field.
- They should be carefully selected and treated with an organomereurial before planting, to protect the cut-ends from invasion by the fungus.
- Pretreating the setts with hot water has been found to stimulate germination of buds and hasten growth so as to help the young plants to overcome the competition with the pathogen.
- A certain amount of resistance to the disease has been found in a few sugarcane varieties in other countries, but field sanitation practices combined with chemical pre-treatment of the setts are quite effective in controlling the disease.

Wilt

Causal Organism: Cephalosporium sacchari

Class: Deuteromycetes

Order: Moniliales

Family: Moniliacease

- This is one of the early known diseases of sugarcane in India.
- It was first reported by Butler and Khan in 1913, from North India.
- It has been reported to cause severe damage to sugarcane crops in many parts of India. During 1965-1967 it caused severe damage to sugarcane crop in the Deccan plateau
- Same disease has been reported from Mexico, Argentina, Barbados, Columbia, Trinidad, Uganda, South Africa and United States of America.

Symptoms

- The first symptoms of the disease become apparent only when the plant has grown for about 4-5 months.
- The canes show gradual withering.
- On examination of affected clumps, the pith will be seen discoloured purple or dirty reef, with longitudinal streaks
- The leaves of affected clumps gradually turn yellow and dry up.
- A characteristic disagreeable odour is also associated with such diseased canes.
- A cottony white mycelium can also be seen in the pith region.
- Frequently this fungal disease is associated with a saprophytic bacterial growth and often the bacteria are mistaken as causal agents.

Disease Cycle

- The fungal mycelium is abundant in the infected canes.
- The hyphae are hyaline, thin walled and septate.
- They produce numerous microconidia on simple or branched, lateral or terminal hyphae, but NO macroconidia are produced.
- This is an important character which is distint from that of Fusarium. The conidia are oval to elliptical, and measure 4-12 x 2-3μm in size. They are mostly unicellular, but the ones formed later in the advanced growth of the fungus may be septate
- Conidia readily germinate to produce single germ tubes.
- The fungus is transmitted from place to place through the infected seed setts.
- When the diseased setts are planted, the eyes may fail to develop or often the shoots arising from the eyes may wilt, due to the infection spreading to the shoots.
- Root formation in such setts may be very poor. The fungus can also survive in soil as a saprophyte for 2-3 years.
- Near-neutral and alkaline soils are favoured by the fungus. The perfect stage is not known.

Control

1. The disease is controlled by selecting seed setts from disease free areas.

2. Alkaline soils may be avoided for growing the crop. The setts, selected from disease free stalk, should be dipped in organomercurial fungicide before planting.
3. Dipping the setts in 40 ppm of boron or manganese, or spraying the plants with either of these minor elements reduces the disease intensity.
4. The varieties Co-617 and B.P.-17 are more resistant than others; hence they should be used when growing crop in the wilt-sick soils.

Grassy Shoot

- This disease was first observed in Maharashtra on sugarcane variety Co. 419 in 1942.
- At present it is reported to be widespread in parts of Andhra Pradesh, Mysore, Bihar, U.P. and Tamil Nadu.
- The disease is more severe on the ratoon crops than on the first planted ones.

Symptoms

- The disease is characterized by proliferation of vegetative buds from the base of the cane giving rise to crowded bunch of tillers bearing narrow leaves.
- The tillers bear pale yellow to completely chlorotic leaves.
- Cane formation rarely takes place in affected clumps and if formed the canes are thin with short internodes.
- The virus is readily transmitted by sap inoculation and in the field it is transmitted through infected setts and perpetuated through crop ratooning. The aphids are the vectors for this disease.
- The same virus also infects jowar, Napier grass and Madras grass.

Control

- The disease is controlled by eradication of diseased parts as soon as symptoms are noticed
- Avoid selection of setts from diseased area
- pre-treating the healthy setts with hot water at 52°C for 1 hour before planting
- Treating them with hot air at 54°C for 8 hours and spraying twice a month with aphidicides.

Gummosis or Gumming Disease

Causal Organism: Xanthomonas vasculorum

Class: Schizomycetes

Order: Pseudomonadales

Family: Pseudomonadaceae

- This is one of the oldest and most common diseases of sugarcane known in many counties.
- The disease was first reported from Brazil in 1869, since when it has been reported from Puerto Rico, West Indies, Columbia and Australia.
- In India Rangaswami reported its occurrence in Tamil Nadu State in 1960, and it is now known in other sugarcane areas of South India.

Symptoms

- The disease is characterized by three distinct symptoms.
- On the leaves characteristic longitudinal streaks or stripes, 1/ 8 to ¼ inch in width and several inches in length are found, these stripes are pale yellow in colour, later turning brown.
- The affected canes are stunted with short internodes, giving the plant a bushy appearance. When such canes are split open or cut transversely, a dull yellow bacterial ooze comes from the cut ends.
- In advanced cases, cavities may develop in the pith region.
- The fibro-vascular bundles are deep red and this is more intense at the nodal regions of the stem.
- The causal bacterium is variously named by different authors, but the above name is universally accepted at present.
- The bacterium is a hort rod, 1.0-11.5μm x 0.4-0.5μm, motile by means of a single polar flagellum.
- Gram negative, non-spore forming, non-capsular and non-acid fast.
- On beer extract agar it produces a yellow spreading and slimy growth.

Disease Cycle

- The disease is primarily spread through the setts taken from diseased plants.

- The secondary spread may be through agricultural implements, including the cutting knife.
- In the field the infection may spread by wind and water, the source of pathogen being the gummy bacterial ooze coming out of diseased tissues.
- Certain insects, particularly flies, play a significant role in transmitting the pathogen from place to place.
- The bacterium can survive in the insect's body for a long time and in this way may be transmitted long distances.
- On entry into the host the bacterium reaches the vascular tissues and becomes systemic.
- Besides sugarcane, several plant species falling under Graminae, including maize, sorghum, Panicum spp. and some other grasses have been found to be infected by this bacterium.
- These hosts may help in many ways the bacterial spread and perpetuation.

Control

- As in other diseases of sugarcane, gummosis can be controlled by selecting setts from disease-free plants, and adopting strict field sanitation practices.
- In the Fiji Island the disease was completely eradicated by destroying the diseased canes and growing resistant varieties.
- Studies on the availability of resistant stock among the sugarcane varieties in India are needed.

Foliar Diseases

- Pokkah Boeg caused by Gibberella fujikuroi (Saw.) Wollenw. Wineland with its imperfect stage Fusarium moniliforme Sheldon.
- Downy mildew caused by Sclerospora sacchari Miyake.
- Eye spot caused by Helminthosporium sacchari (Breda de Haan) Butl.
- Yellow spot caused by Cercospora Kopkei Krueg.
- Brown spot caused by Cercospora Longipes Butl.
- Leaf spot caused by Cercospora vaginae Krueg.
- Rust caused by Puccinia kuchnii (Kr.) Butl. and P.erianthi Padwick & Khan.
- Ring spot caused by Letophaeria sacchari Breda de Haan.

- Leaf blotch or spot caused by Helminthosporium halodes Drechsler, H. sacchari and H. tetramera McKinney.
- Root rot caused by Pythium spp.
- Smut caused by Sphacelotheca erianthi.
- Ergot and sugary disease caused by Claviceps spp.
- Root rot caused by Rhizoctonia solani Kuhn and Sclerotium rolfsii Sacc. The symptoms of these diseases are similar to the ones caused by species of the same genus on other hosts.

Root-Knot Nematode

Casual Organism: Meloidogyne javanica

Class: Nematoda

Order: Tylenchida

Family: Heteroderidae

- Over 50 nematode species belonging to 20 different genera are reported to infect sugarcane in various countries of the world.
- At least 10 have been reported in India.
- Among these the Meloidogyme spp. are important.
- Sugarcane is among the various hosts affected by nematodes in India.
- The crop is damaged by the root-knot nematodes, though several other ecto-parasitic nematodes occur on this host.
- The occurrence of root-knot and other hosts was first established by Rangaswami and his associates during 1958-1960.

Symptoms

- The diseased plants are chlorotic and stunted, and yellow stripes show on the young leaves while older leaves appear healthy.
- Crops at all stages of growth are affected and the symptoms are more prominent during summer months.
- When the roots of affected plants are dug out and examined, they are found to be knotted.
- The young, white roots show much less knotting than the older, wiry ones.
- The knots are usually linear and are found more commonly towards the root tips. They are about 5-8 mm in thickness.

- The nematodes are separated from the infected tissues by the Baermann Funnel technique and examined under the microscope.
- The females measure 400-460mm in length and 45-50mm at the thickest point, oesophagus 40-45mm and tail 30-35mm.
- The males measures 420-460mm in length, 20-25mm in width at the thickest point and oesophagus 55-60mm and tail 30-35mm.
- The cyst is pear shaped, measuring 600-650mm x 380-400mm with a prominent beak.
- The other species of Meloidogyne reported on sugarcane are M. arenaria Chitwood and M. incognita Chitwood, both of lesser importance than M. javanica.

Disease Cycle

- The nematode eggs persist in soil in cyst form and when the susceptible host is planted, the cyst give rise to the larval forms which invade the root and form the galls.
- There are reports on the exudation of specific chemical substances by the various hosts with either attract or repulse the nematode larvae.
- Once the larvae are inside the host tissue, they feed through the vascular bundles and complete the life cycle.
- They produce the cysts, which protrude from the host roots and remain there until harvest.
- The production of growth regulators such as indole acetic acid in the host tissue is reported to be correlated with the growth promoting and nodule forming effect of the nematodes, as found by Rangaswami and his associates.
- There are over one hundred plant species, belonging to diversified families, which are host plants for M. javanica. No doubt these host plants, particularly the weeds present in sugarcane fields, and the crops used in rotation with sugarcane, play significant roles in the perpetuation and inoculum build-up of the nematode.

Control

- The root-knot nematode can be checked by fumigating the soil with a nematicide such as Nemagon or D.D.T
- The soil must be opened up soon after crop harvest.
- The chemical fumigant should be injected into the soil as per the directions given by the manufacturer after which the soil

is kept covered with a tarpaulin or plastic sheets for a week or two, if necessary.

- The fumes permeate the soil, killing the nematode larvae and cysts, and other organisms as well.
- The soil is safe for planting in 2-3 weeks after fumigation.
- In some countries crop rotation with marigold reduces the nematode population.
- The effect of such plants or their plant roots exudates on the nematode population, and the inter-relationships of other soil microbial populations with the nematode, have to be investigated in detail to understand this complex phenomena has to evolve effective control measures.
- Certain nematophagous fungi like Catenaria vermicola, Arthrobotrys conoides Drech. and Dactylella deodycoides Drech. are present in soils, and this might help in reducing the nematode population.
- Besides Meloidogyne javanica and other species of Meloidogyne which cause root-knot symptoms, Pratylenchus sp., Paratylenchus macrophallus deMan, Trichodorus sp., Rotylenchus sp., Tylenchorhynchus sp, Xiphinema sp., Rotylenchus sp. and Hoplolaimus coronatus Cobb have been reported by Kishan Singh on sugarcane in some parts of India.
- More detailed studies are needed on the disease symptoms, life cycle of the nematodes, nature and estimation of damage and control measures.
- A phanerogamous parasite affecting the roots of sugarcane has also been reported from some parts of India.
- It is identified as Striga euphrasioides Benth.
- This is a partial root parasite, growing up from the roots to form leafy shoots.
- The parasite can synthesize carbohydrates through the green chlorophyll pigments in the leaves but for its other nutrients it depends on the host root.
- It is usually controlled by pulling out the shoots before flowering and seed set.
- Spraying weedicides like 2-4, D will kill the parasite, without affecting the sugarcane plant appreciably.

Ratoon Stunting

Symptoms

- This is a virus diseases of sugarcane found in some parts of India in recent years.
- The affected plants are stunted, the stunting being most severe in stubble and ratoon crops.
- The setts taken from diseased plants germinate poorly and the few shoots that emerge grow very slowly.
- The virus infects many hosts falling under Graminae.
- It is sap-transmissible and no insect vector has been found.

Control

- The disease spreads mainly through planting material, setts should be selected from healthy plants.
- Treating the seets in hot water at 50°C for about 2 hours gives 100 per cent control and has become a regular practice in many countries.

Mosaic

Symptoms

- This disease was first observed in Java in 1892 and since then it has been reported from almost every sugarcane growing tract of the world.
- The disease carries different names in different countries but the most common name is 'sugarcane mosaic.'
- During 1919-1920 it caused devastating damage to the sugar industry in Puerto Rico and Cuba, and this led to concentrated studies to establish the causal agent, which was uncertain until about 1920.
- In India the disease was identified during the early thirties of this century and has been reported since then from almost all the sugarcane tracts of the country.
- The intensity of the disease and the loss in crop yield vary considerably, depending upon the host variety and climatic conditions.

Causal Organism

- Sugarcane Mosaic Virus (Marmor sacchari).
- The virus is present in abundance in the chlorotic lesions of the affected leaves.

Bibliography

Aaboe, Asger: *Episodes from the Early History of Mathematics*, New York, 1964.

Andrews, J.: *The Domesticated Capsicums*, University of Texas Press. 1995.

Andrews, L.: *Citrus Production - Orange*, St. Augustine, Trinidad and Tobago, 1990

Bag, A. K.: *Mathematics in Ancient and Medieval India*, Varanasi: Chaukhambha Orientalia, 1979.

Banki, L.: *Bioassay of Pesticides in the Laboratory,* Akademiai Kiado, Budapest, 1978.

Baruah, Akhil: *Advanced Morphology of Angiosperms*, Aavishkar, Delhi, 2008.

Chambers, R.: *Rural Development: Putting the Last First*, London, Longman, 1983.

Clawson, Calvin: *The Mathematical Traveller*, New York, Plenum Press, 1994.

Crump, Thomas: *The Anthropology of Numbers*, Cambridge, Cambridge University Press, 1990.

Dabholkar, A.R.: *General Plant Breeding*, Concept, Delhi, 2006.

Dudley, E.: *The Critical Villager: Beyond Community Participation*, London, Routledge, 1993.

Featherly, H. I.: *Taxonomic Terminology of the Higher Plants*, USA, Iowa State College Press, 1954.

Fortuner, R.: *Nematode Identification and Expert System Technology,* New York, Plenum Press, 1988.

Godden, G.: *Growing Citrus Trees*, Australia, Lothian Publishing Company Pvt. Ltd., 1988.

Gowen, S.: *Banana and Plantains*, London, Chapman Hall, 1996.

Herklots, G. A. C.: *Vegetables in South East Asia*, London, George Allen & Unwin Ltd., 1972.

Huaman, Z.: *Descriptors for Sweet Potato*, Rome, International Board for Plant Genetic Resources, 1991.

Jeffers, P.: *Evaluation of Four Onion Varieties in Montserrat*, Plymouth, CARDI, 1992.

Jones, R. M.: *Plant Resources of South-East Asia,* Wageningen, Pudoc Scientific Publishers, 1992.

Julia, F.: *Fruits of Warm Climates*, Miami, Julia F. Morton Publisher, 1987.

Kim, J. S.: *Sources of Resistance to Viruses of Pepper (Capsicum spp.): a Catalogue*, Taipei, AVRDC, 1994.

Kroll, R.: *Cut Flowers*, Wageningen, CTA, 1995.

Lorenzen, S.: *The Phylogenetic Systematics of Freeliving Nematodes*, London, The Ray Society, 1994.

Macself, A.J.: *Soils and Fertilizers*, Satish Serial Pub, Delhi, 2005.

Malins, A.: *Postharvest Handling of Pineapple and Mango*, Port of Spain, Trinidad and Tobago, 1992.

Mellor, J. W.: *The New Economics of Growth*, Ithaca, Cornell University Press, 1976.

Muneesh Kainth: *Chordate Embryology*, Dominant, Delhi, 2003.

Nobel, P. S.: *Physicochemical and Environmental Plant Physiology*, Academic Press, San Diego, 1999.

Oldham, P.: *Cost of Production of Major Tree Crops in Dominica*, Roseau, Ministry of Agriculture, 1991.

Parry, M.L.: *Climatic Change, Agriculture and Settlements*, Dawson Folkestone UK, 1978.

Pemberton, R. W.: *Predictable Risk to Native Plants in Weed Biological Control,* Oecologia, 2000.

Qystein, V. Sjaastad: *Physiology of Domestic Animals*, International Book Distributing Co., Delhi, 2005.

Ragone, D.: *Breadfruit: Artocarpus Altilis (Parkinson) Fosberg*, Rome, International Plant Genetic Resources Institute, 1997.

Sanchez, P. A.: *Properties and Management of Soils in the Tropics*, John Wiley and Sons, New York, 1976.

Steferud, Alfred: *Diseases of Fruits and Nuts*, Biotech Books, Delhi, 2005.

Tawde, A. B.: *Propagation and Rootstocks of Mango*, New Delhi, Malhotra, 1993.

Urton, Gary: *The Social Life of Numbers*, Austin, University of Texas Press, 1997.

Vanderplank, J.E.: *Plant Diseases: Epidemics and Control,* New York, NY, USA, Academic Press, 1963.

White, G.F.: *Natural Hazards: Local, National, Global*, Oxford University Press, New York, 1974.

Yoshida, T.: *Cultivation of Citrus Genetic Resources for Evaluation of Characteristics*, Tokyo, JICA, 1996.

Index

M

N

P

Q

R

S

T

V

W

❑❑❑